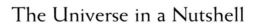

The Universe in a Nutshell

The Universe in a Nutshell

Stephen Hawking

Bantam Books

New York   Toronto   London   Sydney   Auckland

CONTENTS

Stephen Hawking in 2001, © Stewart Cohen.

FOREWORD

I HADN'T EXPECTED MY POPULAR BOOK, *A Brief History of Time*, to be such a success. It was on the London *Sunday Times* bestseller list for over four years, which is longer than any other book has been, and remarkable for a book on science that was not easy going. After that, people kept asking when I would write a sequel. I resisted because I didn't want to write *Son of Brief History* or *A Slightly Longer History of Time*, and because I was busy with research. But I have come to realize that there is room for a different kind of book that might be easier to understand. *A Brief History of Time* was organized in a linear fashion, with most chapters following and logically depending on the preceding chapters. This appealed to some readers, but others got stuck in the early chapters and never reached the more exciting material later on. By contrast, the present book is more like a tree: Chapters 1 and 2 form a central trunk from which the other chapters branch off.

The branches are fairly independent of each other and can be tackled in any order after the central trunk. They correspond to areas I have worked on or thought about since the publication of *A Brief History of Time*. Thus they present a picture of some of the most active fields of current research. Within each chapter I have also tried to avoid a single linear structure. The illustrations and their captions provide an alternative route to the text, as in *The Illustrated Brief History of Time*, published in 1996; and the boxes, or sidebars, provide the opportunity to delve into certain topics in more detail than is possible in the main text.

In 1988, when *A Brief History of Time* was first published, the ultimate Theory of Everything seemed to be just over the horizon. How has the situation changed since then? Are we any closer to our goal? As will be described in this book, we have advanced a long way since then. But it is an ongoing journey still and the end is not yet in sight. According to the old saying, it is better to travel hopefully than to arrive. Our quest for discovery fuels our creativity in all fields, not just science. If we reached the end of the line, the human spirit would shrivel and die. But I don't think we will ever stand still: we shall increase in complexity, if not in depth, and shall always be the center of an expanding horizon of possibilities.

I want to share my excitement at the discoveries that are being made and the picture of reality that is emerging. I have concentrated on areas I have worked on myself for a greater feeling of immediacy. The details of the work are very technical but I believe the broad ideas can be conveyed without a lot of mathematical baggage. I just hope I have succeeded.

I have had a lot of help with this book. I would mention in particular Thomas Hertog and Neel Shearer, for assistance with the figures, captions, and boxes, Ann Harris and Kitty Ferguson, who edited the manuscript (or, more accurately, the computer files, because everything I write is electronic), Philip Dunn of the Book Laboratory and Moonrunner Design, who created the illustrations. But beyond that, I want to thank all those who have made it possible for me to lead a fairly normal life and carry on scientific research. Without them this book could not have been written.

Stephen Hawking
Cambridge, May 2, 2001.

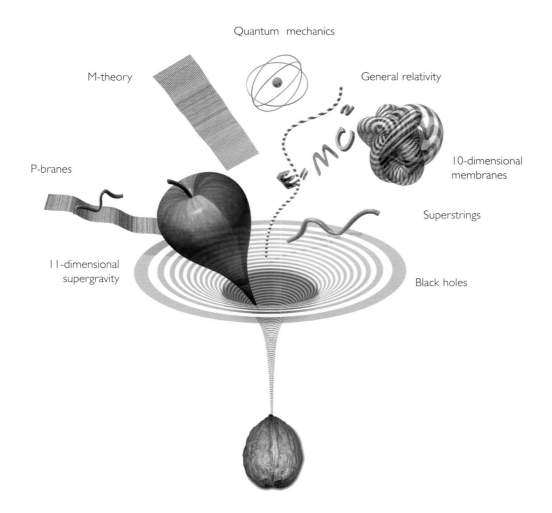

Quantum mechanics

M-theory

General relativity

$E=mc^2$

P-branes

10-dimensional membranes

Superstrings

11-dimensional supergravity

Black holes

CHAPTER 1

A BRIEF HISTORY OF RELATIVITY

*How Einstein laid the foundations of the two fundamental theories
of the twentieth century: general relativity and quantum theory.*

Professor Einstein

Albert Einstein™

ALBERT EINSTEIN, THE DISCOVERER OF THE SPECIAL AND general theories of relativity, was born in Ulm, Germany, in 1879, but the following year the family moved to Munich, where his father, Hermann, and uncle, Jakob, set up a small and not very successful electrical business. Albert was no child prodigy, but claims that he did poorly at school seem to be an exaggeration. In 1894 his father's business failed and the family moved to Milan. His parents decided he should stay behind to finish school, but he did not like its authoritarianism, and within months he left to join his family in Italy. He later completed his education in Zurich, graduating from the prestigious Federal Polytechnical School, known as the ETH, in 1900. His argumentative nature and dislike of authority did not endear him to the professors at the ETH and none of them offered him the position of assistant, which was the normal route to an academic career. Two years later, he finally managed to get a junior post at the Swiss patent office in Bern. It was while he held this job that in 1905 he wrote three papers that both established him as one of the world's leading scientists and started two conceptual revolutions—revolutions that changed our understanding of time, space, and reality itself.

Toward the end of the nineteenth century, scientists believed they were close to a complete description of the universe. They imagined that space was filled by a continuous medium called the "ether." Light rays and radio signals were waves in this ether, just as sound is pressure waves in air. All that was needed for a complete theory were careful measurements of the elastic properties of the ether. In fact, anticipating such measurements, the Jefferson Lab at Harvard University was built entirely without iron nails so as not to interfere with delicate magnetic measurements. However, the planners forgot that the reddish brown bricks of which the lab and most of Harvard are built contain large amounts of iron. The building is still in use today, although Harvard is still not sure how much weight a library floor without iron nails will support.

Albert Einstein™

Albert Einstein in 1920.

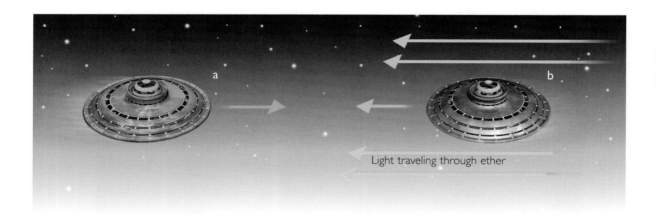

a   b

Light traveling through ether

(FIG. 1.1, above)

THE FIXED ETHER THEORY

If light were a wave in an elastic material called ether, the speed of light should appear higher to someone on a spaceship **(a)** moving toward it, and lower on a spaceship **(b)** traveling in the same direction as the light.

(FIG. 1.2, opposite)

No difference was found between the speed of light in the direction of the Earth's orbit and in a direction at right angles to it.

By the century's end, discrepancies in the idea of an all-pervading ether began to appear. It was expected that light would travel at a fixed speed through the ether but that if you were traveling through the ether in the same direction as the light, its speed would appear lower, and if you were traveling in the opposite direction of the light, its speed would appear higher (Fig. 1.1).

Yet a series of experiments failed to support this idea. The most careful and accurate of these experiments was carried out by Albert Michelson and Edward Morley at the Case School of Applied Science in Cleveland, Ohio, in 1887. They compared the speed of light in two beams at right angles to each other. As the Earth rotates on its axis and orbits the Sun, the apparatus moves through the ether with varying speed and direction (Fig. 1.2). But Michelson and Morley found no daily or yearly differences between the two beams of light. It was as if light always traveled at the same speed relative to where one was, no matter how fast and in which direction one was moving (Fig. 1.3, page 8).

Based on the Michelson-Morley experiment, the Irish physicist George FitzGerald and the Dutch physicist Hendrik Lorentz suggested that bodies moving through the ether would contract and that clocks would slow down. This contraction and the slowing down of clocks would be such that people would all measure the same speed for light, no matter how they were moving with respect to the ether. (FitzGerald and Lorentz still regarded ether as a real substance.) However, in a paper written in June 1905, Einstein

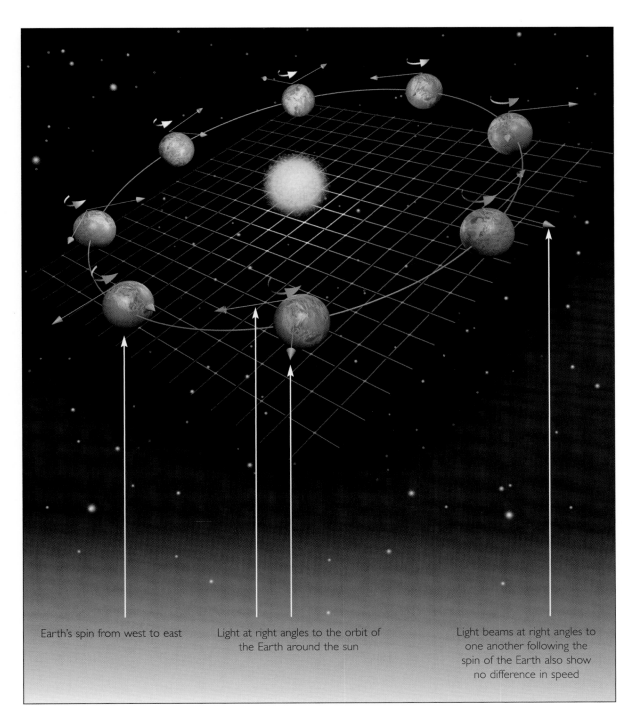

Earth's spin from west to east

Light at right angles to the orbit of
the Earth around the sun

Light beams at right angles to
one another following the
spin of the Earth also show
no difference in speed

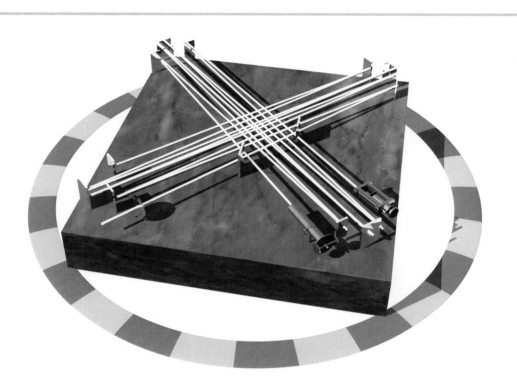

(FIG. 1.3) MEASURING THE SPEED OF LIGHT

In the Michelson-Morley interferometer, light from a source is split into two beams by a half-silvered mirror. The two beams of light travel at right angles to each other and are then combined into a single beam by hitting the half-silvered mirror again. A difference in the speed of light traveling in the two directions could mean that the wave crests in one beam arrived at the same time as the wave troughs of the other and canceled them out.

Right: Diagram of the experiment reconstructed from that which appeared in the *Scientific American* of 1887.

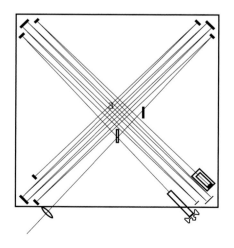

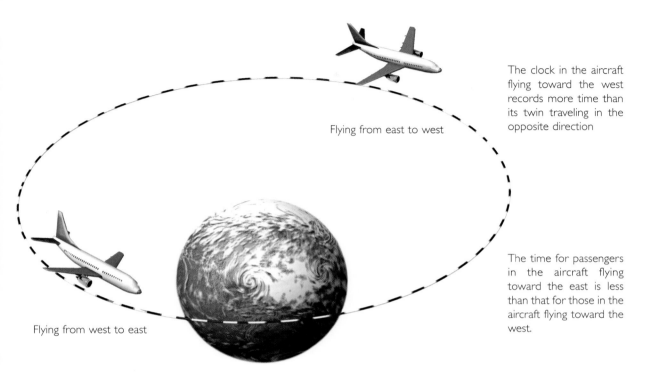

Flying from east to west

The clock in the aircraft flying toward the west records more time than its twin traveling in the opposite direction

Flying from west to east

The time for passengers in the aircraft flying toward the east is less than that for those in the aircraft flying toward the west.

pointed out that if one could not detect whether or not one was moving through space, the notion of an ether was redundant. Instead, he started from the postulate that the laws of science should appear the same to all freely moving observers. In particular, they should all measure the same speed for light, no matter how fast they were moving. The speed of light is independent of their motion and is the same in all directions.

This required abandoning the idea that there is a universal quantity called time that all clocks would measure. Instead, everyone would have his or her own personal time. The times of two people would agree if the people were at rest with respect to each other, but not if they were moving.

This has been confirmed by a number of experiments, including one in which two accurate clocks were flown in opposite directions around the world and returned showing very slightly different times (Fig. 1.4). This might suggest that if one wanted to live longer, one should keep flying to the east so that the plane's speed is added to the earth's rotation. However, the tiny fraction of a second one would gain would be more than canceled by eating airline meals.

(FIG. 1.4)
One version of the twins paradox (Fig. 1.5, page 10) has been tested experimentally by flying two accurate clocks in opposite directions around the world.

When they met up again the clock that flew toward the east had recorded slightly less time.

(FIG. 1.5, left.)
THE TWINS PARADOX

In the theory of relativity each observer has his own measure of time. This can lead to the so-called twins paradox.

One of a pair of twins (a) leaves on a space journey during which he travels close to the speed of light (c), while his brother (b) remains on Earth.
    Because of (a)'s motion, time runs more slowly in the spacecraft as seen by the earthbound twin. So on his return the space traveler (a2) will find that his brother (b2) has aged more than himself.
    Although it seems against common sense, a number of experiments have implied that in this scenario the traveling twin would indeed be younger.

(FIG. 1.6, right).
A spaceship passes Earth from left to right at four-fifths the speed of light. A pulse of light is emitted at one end of the cabin and reflected at the other end (a).
    The light is observed by people on Earth and on the spaceship. Because of the motion of the spaceship, they will disagree about the distance the light has traveled in reflecting back (b).
    They must therefore also disagree about the time the light has taken, because according to Einstein's postulate the speed of light is the same for all freely moving observers.

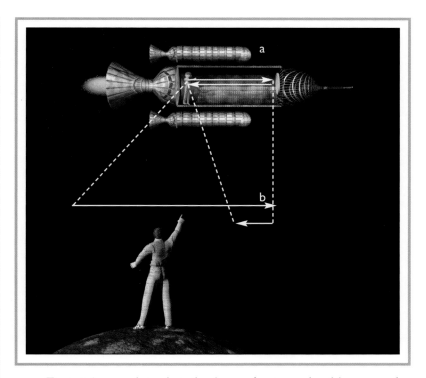

Einstein's postulate that the laws of nature should appear the same to all freely moving observers was the foundation of the theory of relativity, so called because it implied that only relative motion was important. Its beauty and simplicity convinced many thinkers, but there remained a lot of opposition. Einstein had overthrown two of the absolutes of nineteenth-century science: absolute rest, as represented by the ether, and absolute or universal time that all clocks would measure. Many people found this an unsettling concept. Did it imply, they asked, that *everything* was relative, that there were no absolute moral standards? This unease continued throughout the 1920s and 1930s. When Einstein was awarded the Nobel Prize in 1921, the citation was for important but (by his standard) comparatively minor work also carried out in 1905. It made no mention of relativity, which was considered too controversial. (I still get two or three letters a week telling me Einstein was wrong.) Nevertheless, the theory of relativity is now completely accepted by the scientific community, and its predictions have been verified in countless applications.

FIG. 1.7

A very important consequence of relativity is the relation between mass and energy. Einstein's postulate that the speed of light should appear the same to everyone implied that nothing could be moving faster than light. What happens is that as one uses energy to accelerate anything, whether a particle or a spaceship, its mass increases, making it harder to accelerate it further. To accelerate a particle to the speed of light would be impossible because it would take an infinite amount of energy. Mass and energy are equivalent, as is summed up in Einstein's famous equation $E = mc^2$ (Fig. 1.7). This is probably the only equation in physics to have recognition on the street. Among its consequences was the realization that if the nucleus of a uranium atom fissions into two nuclei with slightly less total mass, this will release a tremendous amount of energy (see pages 14-15, Fig. 1.8).

In 1939, as the prospect of another world war loomed, a group of scientists who realized these implications persuaded Einstein to overcome his pacifist scruples and add his authority to a letter to

EINSTEIN'S PROPHETIC LETTER TO
PRESIDENT ROOSEVELT IN 1939

"In the course of the
last four months it
has been made probable
—through the work of
Joliot in France as
well as Fermi and
Szilard in America—
that it may become
possible to set up a
nuclear chain reaction
in a large mass of
uranium, by which vast
amounts of power and
large quantities of new
radium-like elements
would be generated.
Now it appears almost
certain that this could
be achieved in the
immediate future.
  This new phenomenon
would also lead to the
construction of bombs,
and it is conceivable
—though much less
certain—that extremely
powerful bombs of a
new type may thus be
constructed."

President Roosevelt urging the United States to start a program of nuclear research.

This led to the Manhattan Project and ultimately to the bombs that exploded over Hiroshima and Nagasaki in 1945. Some people have blamed the atom bomb on Einstein because he discovered the relationship between mass and energy; but that is like blaming Newton for causing airplanes to crash because he discovered gravity. Einstein himself took no part in the Manhattan Project and was horrified by the dropping of the bomb.

After his groundbreaking papers in 1905, Einstein's scientific reputation was established. But it was not until 1909 that he was offered a position at the University of Zurich that enabled him to leave the Swiss patent office. Two years later, he moved to the German University in Prague, but he came back to Zurich in 1912, this time to the ETH. Despite the anti-Semitism that was common in much of Europe, even in the universities, he was now an academic hot property. Offers came in from Vienna and Utrecht, but he chose to

Uranium (**U-235**)

Uranium (**U-236**)

(**n**)

Gamma ray

(**n**)

Impact by neutron (**n**)

(**U-235**) compound nucleus oscillates and is unstable

(**Ba-144**) compound nucleus oscillates and is unstable

(FIG. 1.8)

NUCLEAR BINDING ENERGY

Nuclei are made up of protons and neutrons held together by a strong force. But the mass of the nucleus is always less than the sum of the individual masses of the protons and neutrons that make it up. The difference is a measure of the nuclear binding energy that holds the nucleus together. This binding energy can be calculated from the Einstein relationship: nuclear binding energy $= \Delta \mathbf{mc^2}$ where $\Delta\mathbf{m}$ is the difference between the mass of the nucleus and the sum of the individual masses.

It is the release of this potential energy that creates the devastating explosive force of a nuclear device.

accept a research position with the Prussian Academy of Sciences in Berlin because it freed him from teaching duties. He moved to Berlin in April 1914 and was joined shortly after by his wife and two sons. The marriage had been in a bad way for some time, however, and his family soon returned to Zurich. Although he visited them occasionally, he and his wife were eventually divorced. Einstein later married his cousin Elsa, who lived in Berlin. The fact that he spent the war years as a bachelor, without domestic commitments, may be one reason why this period was so productive for him scientifically.

Although the theory of relativity fit well with the laws that governed electricity and magnetism, it was not compatible with Newton's law of gravity. This law said that if one changed the distribution of matter in one region of space, the change in the gravitational field would be felt instantaneously everywhere else in the universe. Not only would this mean one could send signals faster than light (something that was forbidden by relativity); in order to know what instantaneous meant, it also required the existence of absolute or universal time, which relativity had abolished in favor of personal time.

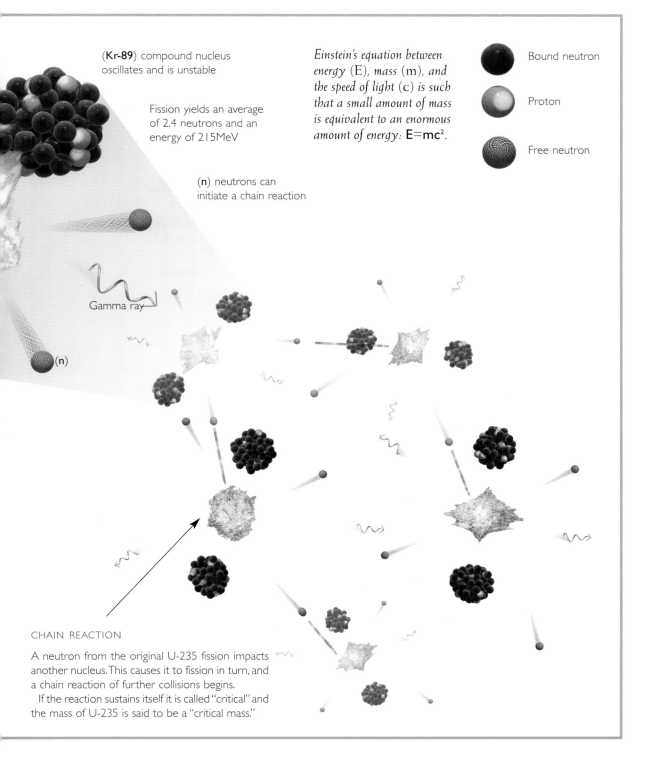

(**Kr-89**) compound nucleus oscillates and is unstable

Fission yields an average of 2.4 neutrons and an energy of 215MeV

Einstein's equation between energy (E), *mass* (m), *and the speed of light* (c) *is such that a small amount of mass is equivalent to an enormous amount of energy:* $E=mc^2$.

Bound neutron

Proton

Free neutron

(**n**) neutrons can initiate a chain reaction

Gamma ray

(n)

CHAIN REACTION

A neutron from the original U-235 fission impacts another nucleus. This causes it to fission in turn, and a chain reaction of further collisions begins.

If the reaction sustains itself it is called "critical" and the mass of U-235 is said to be a "critical mass."

(FIG. 1.9)
An observer in a box cannot tell the difference between being in a stationary elevator on Earth (a) and being accelerated by a rocket in free space (b).

If the rocket motor is turned off (c), it feels as if the elevator is in free fall to the bottom of the shaft (d).

Einstein was aware of this difficulty in 1907, while he was still at the patent office in Bern, but it was not until he was in Prague in 1911 that he began to think seriously about the problem. He realized that there is a close relationship between acceleration and a gravitational field. Someone inside a closed box, such as an elevator, could not tell whether the box was at rest in the Earth's gravitational field or was being accelerated by a rocket in free space. (Of course, this was before the age of *Star Trek*, and so Einstein thought of people in elevators rather than spaceships.) But one cannot accelerate or fall freely very far in an elevator before disaster strikes (Fig. 1.9).

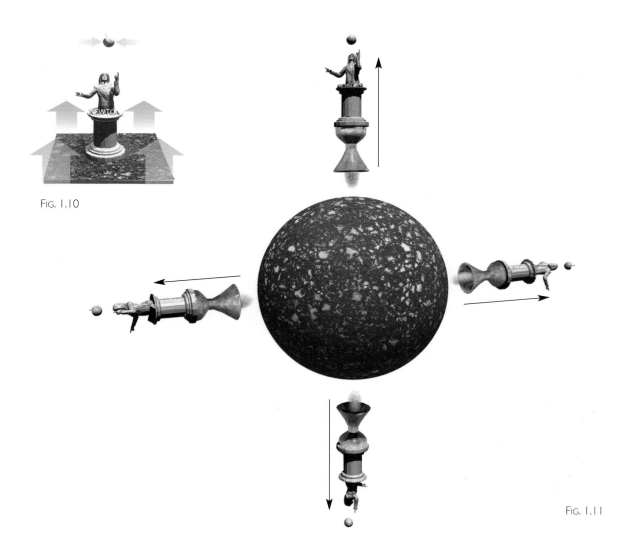

Fig. 1.10

Fig. 1.11

If the Earth were flat, one could equally well say that the apple fell on Newton's head because of gravity or because Newton and the surface of the Earth were accelerating upward (Fig. 1.10). This equivalence between acceleration and gravity didn't seem to work for a round Earth, however—people on the opposite sides of the world would have to be accelerating in opposite directions but staying at a constant distance from each other (Fig. 1.11).

But on his return to Zurich in 1912 Einstein had the brain wave of realizing that the equivalence would work if the geometry of spacetime was curved and not flat, as had been assumed hitherto.

If the Earth were flat (FIG. 1.10) one could say that either the apple fell on Newton's head because of gravity or that the Earth and Newton were accelerating upward. This equivalence didn't work for a spherical Earth (FIG. 1.11) because people on opposite sides of the world would be getting farther away from each other. Einstein overcame this difficulty by making space and time curved.

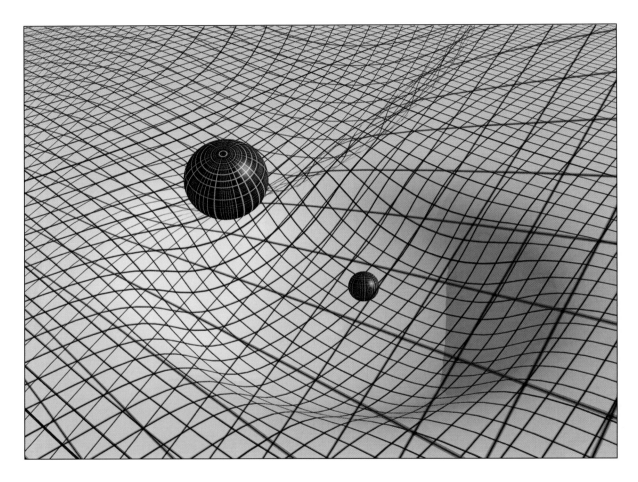

(FIG. 1.12) SPACETIME CURVES

Acceleration and gravity can be equivalent only if a massive body curves spacetime, thereby bending the paths of objects in its neighborhood.

His idea was that mass and energy would warp spacetime in some manner yet to be determined. Objects such as apples or planets would try to move in straight lines through spacetime, but their paths would appear to be bent by a gravitational field because spacetime is curved (Fig. 1.12).

With the help of his friend Marcel Grossmann, Einstein studied the theory of curved spaces and surfaces that had been developed earlier by Georg Friedrich Riemann. However, Riemann thought only of space being curved. It took Einstein to realize that it is spacetime which is curved. Einstein and Grossmann wrote a joint paper in 1913 in which they put forward the idea that what we think of as gravitational forces are just an expression of the fact that

Professor Einstein

spacetime is curved. However, because of a mistake by Einstein (who was quite human and fallible), they weren't able to find the equations that related the curvature of spacetime to the mass and energy in it. Einstein continued to work on the problem in Berlin, undisturbed by domestic matters and largely unaffected by the war, until he finally found the right equations in November 1915. He had discussed his ideas with the mathematician David Hilbert during a visit to the University of Göttingen in the summer of 1915, and Hilbert independently found the same equations a few days before Einstein. Nevertheless, as Hilbert himself admitted, the credit for the new theory belonged to Einstein. It was his idea to relate gravity to the warping of spacetime. It is a tribute to the civilized state of Germany at this period that such scientific discussions and exchanges could go on undisturbed even in wartime. It was a sharp contrast to the Nazi era twenty years later.

The new theory of curved spacetime was called general relativity to distinguish it from the original theory without gravity, which was now known as special relativity. It was confirmed in a spectacular fashion in 1919 when a British expedition to West Africa observed a slight bending of light from a star passing near

Albert Einstein™

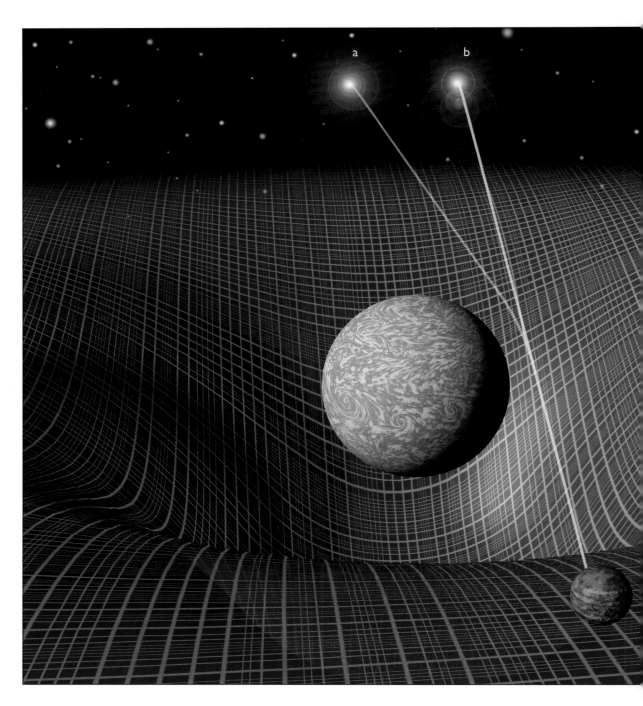

(FIG. 1.13) LIGHT CURVES

Light from a star passing near the Sun is deflected by the way the mass of the Sun curves spacetime (a). This produces a slight shift in the apparent position of the star as seen from the Earth (b). This can be observed during an eclipse.

the sun during an eclipse (Fig. 1.13). Here was direct evidence that space and time are warped, and it spurred the greatest change in our perception of the universe in which we live since Euclid wrote his *Elements of Geometry* around 300 B.C.

Einstein's general theory of relativity transformed space and time from a passive background in which events take place to active participants in the dynamics of the universe. This led to a great problem that remains at the forefront of physics in the twenty-first century. The universe is full of matter, and matter warps spacetime in such a way that bodies fall together. Einstein found that his equations didn't have a solution that described a static universe, unchanging in time. Rather than give up such an everlasting universe, which he and most other people believed in, he fudged the equations by adding a term called the cosmological constant, which warped spacetime in the opposite sense, so that bodies move apart. The repulsive effect of the cosmological constant could balance the attractive effect of the matter, thus allowing a static solution for the universe. This was one of the great missed opportunities of theoretical physics. If Einstein had stuck with his original equations, he could have predicted that the universe must be either expanding or contracting. As it was, the possibility of a time-dependent universe wasn't taken seriously until observations in the 1920s by the 100-inch telescope on Mount Wilson.

These observations revealed that the farther other galaxies are from us, the faster they are moving away. The universe is expanding, with the distance between any two galaxies steadily increasing with time (Fig. 1.14, page 22). This discovery removed the need for a cosmological constant in order to have a static solution for the universe. Einstein later called the cosmological constant the greatest mistake of his life. However, it now seems that it may not have been a mistake after all: recent observations, described in Chapter 3, suggest that there may indeed be a small cosmological constant.

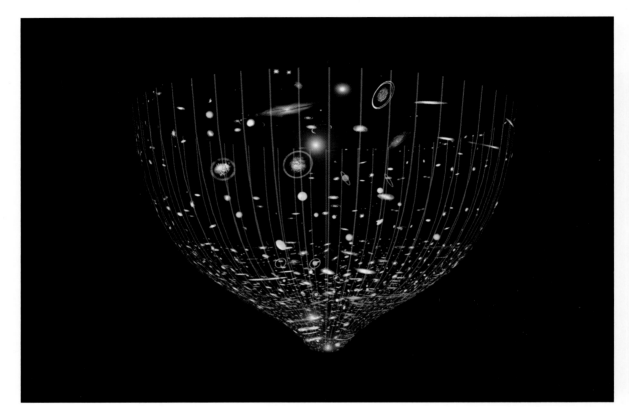

(FIG. 1.14)

Observations of galaxies indicate that the universe is expanding: the distance between almost any pair of galaxies is increasing.

General relativity completely changed the discussion of the origin and fate of the universe. A static universe could have existed forever or could have been created in its present form at some time in the past. However, if galaxies are moving apart now, it means that they must have been closer together in the past. About fifteen billion years ago, they would all have been on top of each other and the density would have been very large. This state was called the "primeval atom" by the Catholic priest Georges Lemaître, who was the first to investigate the origin of the universe that we now call the big bang.

Einstein seems never to have taken the big bang seriously. He apparently thought that the simple model of a uniformly expanding universe would break down if one followed the motions of the galaxies back in time, and that the small sideways velocities of the galaxies would cause them to miss each other. He thought the universe might have had a previous contracting phase, with a bounce into the present expansion at a fairly moderate density. However, we now know that in order for nuclear reactions in the early universe to

The 100-inch Hooker telescope at Mount Wilson Observatory.

produce the amounts of light elements we observe around us, the density must have been at least ten tons per cubic inch and the temperature ten billion degrees. Further, observations of the microwave background indicate that the density was probably once a trillion trillion trillion trillion trillion trillion (1 with 72 zeros after it) tons per cubic inch. We also now know that Einstein's general theory of relativity does not allow the universe to bounce from a contracting phase to the present expansion. As will be discussed in Chapter 2, Roger Penrose and I were able to show that general relativity predicts that the universe began in the big bang. So Einstein's theory does imply that time has a beginning, although he was never happy with the idea.

Einstein was even more reluctant to admit that general relativity predicted that time would come to an end for massive stars when they reached the end of their life and no longer generated enough heat to balance the force of their own gravity, which was trying to make them smaller. Einstein thought that such stars would settle down to some

(FIG. 1.15)
When a massive star exhausts its nuclear fuel, it will lose heat and contract. The warping of spacetime will become so great that a black hole will be created from which light cannot escape. Inside the black hole time will come to an end.

final state, but we now know that there are no final-state configurations for stars of more than twice the mass of the sun. Such stars will continue to shrink until they become black holes, regions of spacetime that are so warped that light cannot escape from them (Fig. 1.15).

Penrose and I showed that general relativity predicted that time would come to an end inside a black hole, both for the star and for any unfortunate astronaut who happened to fall into it. But both the beginning and the end of time would be places where the equations of general relativity could not be defined. Thus the theory could not predict what should emerge from the big bang. Some saw this as an indication of God's freedom to start the universe off in any way God wanted, but others (including myself) felt that the beginning of the universe should be governed by the same laws that held at other times. We have made some progress toward this goal, as will be described in Chapter 3, but we don't yet have a complete understanding of the origin of the universe.

The reason general relativity broke down at the big bang was that it was not compatible with quantum theory, the other great conceptual revolution of the early twentieth century. The first step toward quantum theory had come in 1900, when Max Planck in Berlin discovered that the radiation from a body that was glowing red-hot was explainable if light could be emitted or absorbed only if it came in discrete packets, called quanta. In one of his groundbreaking papers, written in 1905 when he was at the patent office, Einstein showed that Planck's quantum hypothesis could explain what is called the photoelectric effect, the way certain metals give off electrons when light falls on them. This is the basis of modern light detectors and television cameras, and it was for this work that Einstein was awarded the Nobel Prize for physics.

Einstein continued to work on the quantum idea into the 1920s, but he was deeply disturbed by the work of Werner Heisenberg in Copenhagen, Paul Dirac in Cambridge, and Erwin Schrödinger in Zurich, who developed a new picture of reality called quantum mechanics. No longer did tiny particles have a definite position and

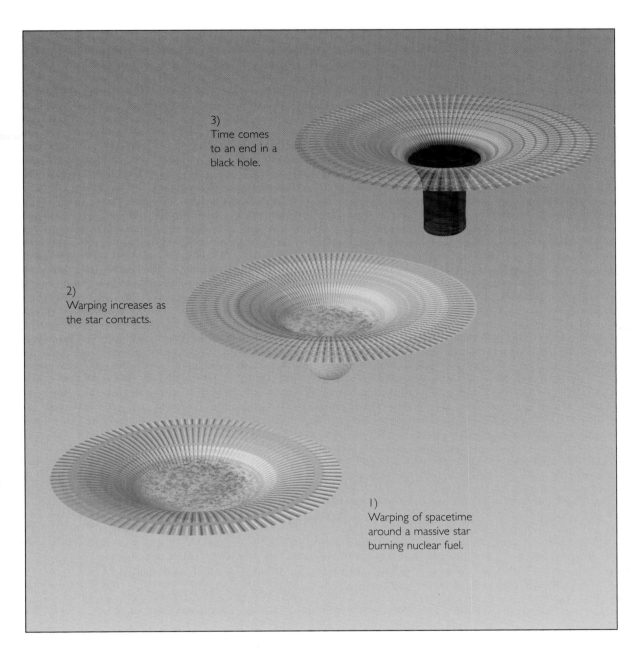

3)
Time comes
to an end in a
black hole.

2)
Warping increases as
the star contracts.

1)
Warping of spacetime
around a massive star
burning nuclear fuel.

Albert Einstein with a puppet of himself shortly after arriving in America for good.

Albert Einstein™

speed. Instead, the more accurately one determined a particle's position, the less accurately one could determine its speed, and vice versa. Einstein was horrified by this random, unpredictable element in the basic laws and never fully accepted quantum mechanics. His feelings were expressed in his famous dictum "God does not play dice." Most other scientists, however, accepted the validity of the new quantum laws because of the explanations they gave for a whole range of previously unaccounted-for phenomena and their excellent agreement with observations. They are the basis of modern developments in chemistry, molecular biology, and electronics, and the foundation for the technology that has transformed the world in the last fifty years.

In December 1932, aware that the Nazis and Hitler were about to come to power, Einstein left Germany and four months later renounced his citizenship, spending the last twenty years of his life at the Institute for Advanced Study in Princeton, New Jersey.

In Germany, the Nazis launched a campaign against "Jewish science" and the many German scientists who were Jews; this is part of the reason that Germany was not able to build an atomic bomb. Einstein and relativity were principal targets of this campaign. When told of the publication of a book entitled 100 *Authors Against Einstein*, he replied: "Why one hundred? If I were wrong, one would have been enough." After the Second World War, he urged the Allies to set up a world government to control the atomic bomb. In 1948, he was offered the presidency of the new state of Israel but turned it down. He once said: "Politics is for the moment, but an equation is for eternity." The Einstein equations of general relativity are his best epitaph and memorial. They should last as long as the universe.

The world has changed far more in the last hundred years than in any previous century. The reason has not been new political or economic doctrines but the vast developments in technology made possible by advances in basic science. Who better symbolizes those advances than Albert Einstein?

CHAPTER 2

THE SHAPE OF TIME

Einstein's general relativity gives time a shape.
How this can be reconciled with quantum theory.

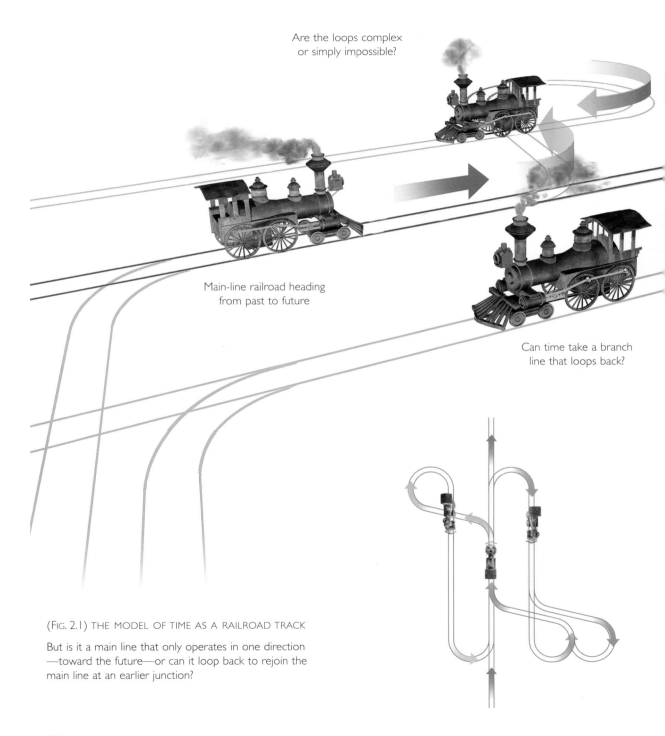

Are the loops complex
or simply impossible?

Main-line railroad heading
from past to future

Can time take a branch
line that loops back?

(FIG. 2.1) THE MODEL OF TIME AS A RAILROAD TRACK

But is it a main line that only operates in one direction
—toward the future—or can it loop back to rejoin the
main line at an earlier junction?

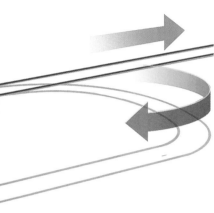

W HAT IS TIME? IS IT AN EVER-ROLLING STREAM THAT bears all our dreams away, as the old hymn says? Or is it a railroad track? Maybe it has loops and branches, so you can keep going forward and yet return to an earlier station on the line (Fig. 2.1).

The nineteenth-century author Charles Lamb wrote: "Nothing puzzles me like time and space. And yet nothing troubles me *less* than time and space, because I never think of them." Most of us don't worry about time and space most of the time, whatever that may be; but we all do wonder sometimes what time is, how it began, and where it is leading us.

Any sound scientific theory, whether of time or of any other concept, should in my opinion be based on the most workable philosophy of science: the positivist approach put forward by Karl Popper and others. According to this way of thinking, a scientific theory is a mathematical model that describes and codifies the observations we make. A good theory will describe a large range of phenomena on the basis of a few simple postulates and will make definite predictions that can be tested. If the predictions agree with the observations, the theory survives that test, though it can never be proved to be correct. On the other hand, if the observations disagree with the predictions, one has to discard or modify the theory. (At least, that is what is supposed to happen. In practice, people often question the accuracy of the observations and the reliability and moral character of those making the observations.) If one takes the positivist position, as I do, one cannot say what time actually is. All one can do is describe what has been found to be a very good mathematical model for time and say what predictions it makes.

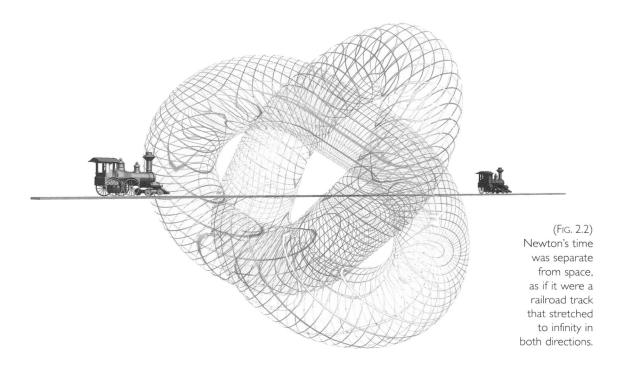

(FIG. 2.2)
Newton's time was separate from space, as if it were a railroad track that stretched to infinity in both directions.

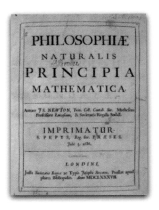

Isaac Newton published his mathematical model of time and space over 300 years ago.

Isaac Newton gave us the first mathematical model for time and space in his *Principia Mathematica*, published in 1687. Newton occupied the Lucasian chair at Cambridge that I now hold, though it wasn't electrically operated in his time. In Newton's model, time and space were a background in which events took place but which weren't affected by them. Time was separate from space and was considered to be a single line, or railroad track, that was infinite in both directions (Fig. 2.2). Time itself was considered eternal, in the sense that it had existed, and would exist, forever. By contrast, most people thought the physical universe had been created more or less in its present state only a few thousand years ago. This worried philosophers such as the German thinker Immanuel Kant. If the universe had indeed been created, why had there been an infinite wait before the creation? On the other hand, if the universe had existed forever, why hadn't everything that was going to happen already happened, meaning that history was over? In particular, why hadn't the universe reached thermal equilibrium, with everything at the same temperature?

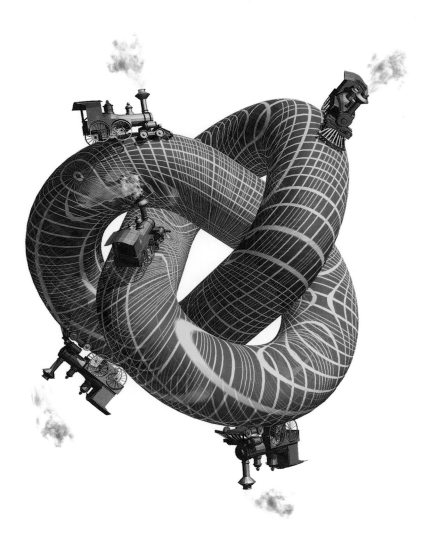

(FIG. 2.3) THE SHAPE AND DIRECTION OF TIME

Einstein's theory of relativity, which agrees with a large number of experiments, shows that time and space are inextricably interconnected.

One cannot curve space without involving time as well. Thus time has a shape. However, it appears to also have a one-way direction, as the locomotives in the illustration show.

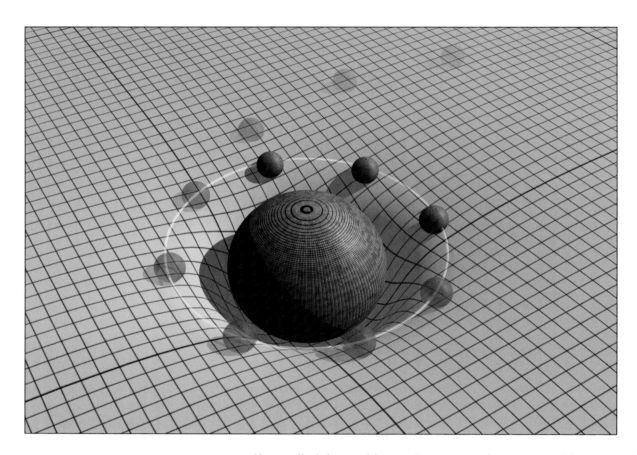

(FIG. 2.4)

THE RUBBER SHEET ANALOGY

The large ball in the center represents a massive body such as a star.

Its weight curves the sheet near it. The ball bearings rolling on the sheet are deflected by this curvature and go around the large ball, in the same way that planets in the gravitational field of a star can orbit it.

Kant called this problem an "antinomy of pure reason," because it seemed to be a logical contradiction; it didn't have a resolution. But it was a contradiction only within the context of the Newtonian mathematical model, in which time was an infinite line, independent of what was happening in the universe. However, as we saw in Chapter 1, in 1915 a completely new mathematical model was put forward by Einstein: the general theory of relativity. In the years since Einstein's paper, we have added a few ribbons and bows, but our model of time and space is still based on what Einstein proposed. This and the following chapters will describe how our ideas have developed in the years since Einstein's revolutionary paper. It has been a success story of the work of a large number of people, and I'm proud to have made a small contribution.

General relativity combines the time dimension with the three dimensions of space to form what is called spacetime (see page 33, Fig. 2.3). The theory incorporates the effect of gravity by saying that the distribution of matter and energy in the universe warps and distorts spacetime, so that it is not flat. Objects in this spacetime try to move in straight lines, but because spacetime is curved, their paths appear bent. They move as if affected by a gravitational field.

As a rough analogy, not to be taken too literally, imagine a sheet of rubber. One can place a large ball on the sheet to represent the Sun. The weight of the ball will depress the sheet and cause it to be curved near the Sun. If one now rolls little ball bearings on the sheet, they won't roll straight across to the other side but instead will go around the heavy weight, like planets orbiting the Sun (Fig. 2.4).

The analogy is incomplete because in it only a two-dimensional section of space (the surface of the rubber sheet) is curved, and time is left undisturbed, as it is in Newtonian theory. However, in the theory of relativity, which agrees with a large number of experiments, time and space are inextricably tangled up. One cannot curve space without involving time as well. Thus time has a shape. By curving space and time, general relativity changes them from being a passive background against which events take place to being active, dynamic participants in what happens. In Newtonian theory, where time existed independently of anything else, one could ask: What did God do before He created the universe? As Saint Augustine said, one should not joke about this, as did a man who said, "He was preparing Hell for those who pry too deep." It is a serious question that people have pondered down the ages. According to Saint Augustine, before God made heaven and earth, He did not make anything at all. In fact, this is very close to modern ideas.

In general relativity, on the other hand, time and space do not exist independently of the universe or of each other. They are defined by measurements within the universe, such as the number of vibrations of a quartz crystal in a clock or the length of a ruler. It is quite conceivable that time defined in this way, within the universe, should have a minimum or maximum value—in other words, a beginning or an end. It would make no sense to ask what happened before the beginning or after the end, because such times would not be defined.

St. Augustine, the fifth-century thinker who held that time did not exist before the beginning of the world.

Page from De Civitate Dei, *twelfth century. Biblioteca Laurenziana, Firenze.*

It was clearly important to decide whether the mathematical model of general relativity *predicted* that the universe, and time itself, should have a beginning or end. The general prejudice among theoretical physicists, including Einstein, held that time should be infinite in both directions. Otherwise, there were awkward questions about the creation of the universe, which seemed to be outside the realm of science. Solutions of the Einstein equations were known in which time had a beginning or end, but these were all very special, with a large amount of symmetry. It was thought that in a real body, collapsing under its own gravity, pressure or sideways velocities would prevent all the matter falling together to the same point, where the density would be infinite. Similarly, if one traced the expansion of the universe back in time, one would find that the matter of the universe didn't all emerge from a point of infinite density. Such a point of infinite density was called a singularity and would be a beginning or an end of time.

In 1963, two Russian scientists, Evgenii Lifshitz and Isaac Khalatnikov, claimed to have proved that solutions of the Einstein equations with a singularity all had a special arrangement of matter and velocities. The chances that the solution representing the universe would have this special arrangement were practically zero. Almost all solutions that could represent the universe would avoid having a singularity of infinite density: Before the era during which the universe has been expanding, there must have been a previous contracting phase during which matter fell together but missed colliding with itself, moving apart again in the present expanding phase. If this were the case, time would continue on forever, from the infinite past to the infinite future.

Not everyone was convinced by the arguments of Lifshitz and Khalatnikov. Instead, Roger Penrose and I adopted a different approach, based not on a detailed study of solutions but on the global structure of spacetime. In general relativity, spacetime is curved not only by massive objects in it but also by the energy in it. Energy is always positive, so it gives spacetime a curvature that bends the paths of light rays toward each other.

Now consider our past light cone (Fig. 2.5), that is, the paths through spacetime of the light rays from distant galaxies that reach

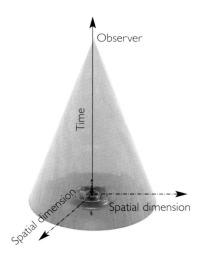

Observer looking back through time

Galaxies as they appeared recently

Galaxies as they appeared 5 billion years ago

The background radiation

Observer

Time

Spatial dimension

Spatial dimension

(FIG. 2.5) OUR PAST LIGHT CONE

When we look at distant galaxies, we are looking at the universe at an earlier time because light travels at a finite speed. If we represent time by the vertical direction and represent two of the three space directions horizontally, the light now reaching us at the point at the top has traveled toward us on a cone.

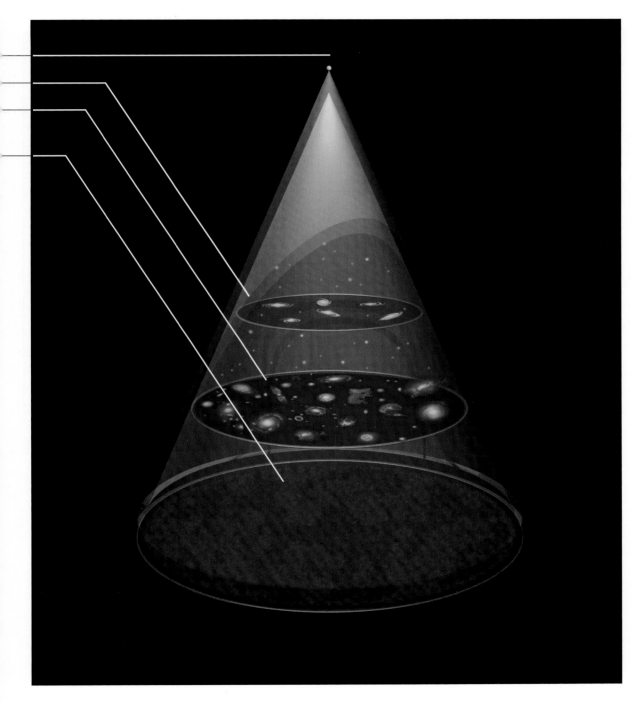

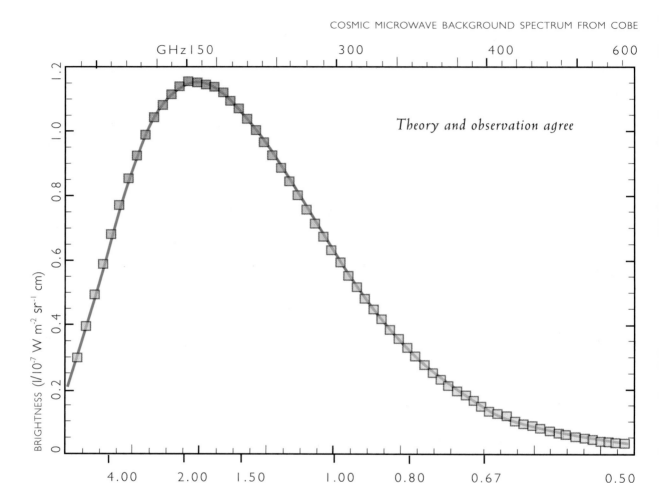

Theory and observation agree

(FIG. 2.6)

MEASUREMENT OF THE SPECTRUM OF MICROWAVE BACKGROUND

The spectrum—the distribution of intensity with frequency—of the cosmic microwave background radiation is characteristic of that from a hot body. For the radiation to be in thermal equilibrium, matter must have scattered it many times. This indicates that there must have been sufficient matter in our past light cone to cause it to bend in.

us at the present time. In a diagram with time plotted upward and space plotted sideways, this is a cone with its vertex, or point, at us. As we go toward the past, down the cone from the vertex, we see galaxies at earlier and earlier times. Because the universe has been expanding and everything used to be much closer together, as we look back further we are looking back through regions of higher matter density. We observe a faint background of microwave radiation that propagates to us along our past light cone from a much earlier time, when the universe was much denser and hotter than it is now. By tuning receivers to different frequencies of microwaves, we can measure the spectrum (the distribution of power arranged

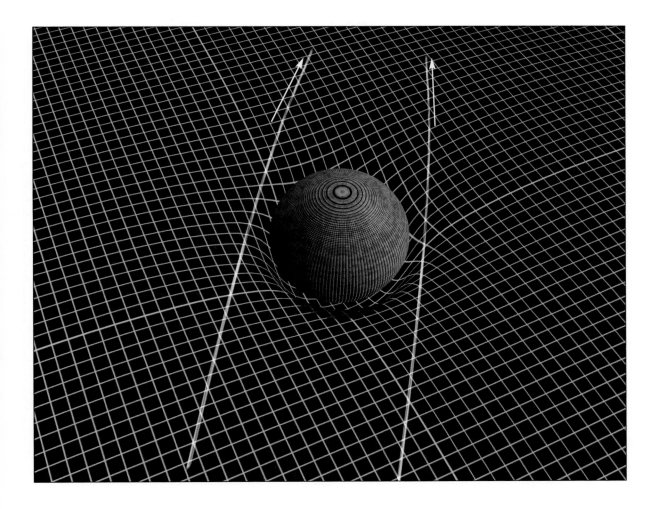

by frequency) of this radiation. We find a spectrum that is characteristic of radiation from a body at a temperature of 2.7 degrees
above absolute zero. This microwave radiation is not much good
for defrosting frozen pizza, but the fact that the spectrum agrees so
exactly with that of radiation from a body at 2.7 degrees tells us that
the radiation must have come from regions that are opaque to
microwaves (Fig. 2.6).

Thus we can conclude that our past light cone must pass
through a certain amount of matter as one follows it back. This
amount of matter is enough to curve spacetime, so the light rays in
our past light cone are bent back toward each other (Fig. 2.7).

(FIG. 2.7) WARPING SPACETIME

Because gravity is attractive, matter
always warps spacetime so that light
rays bend toward each other.

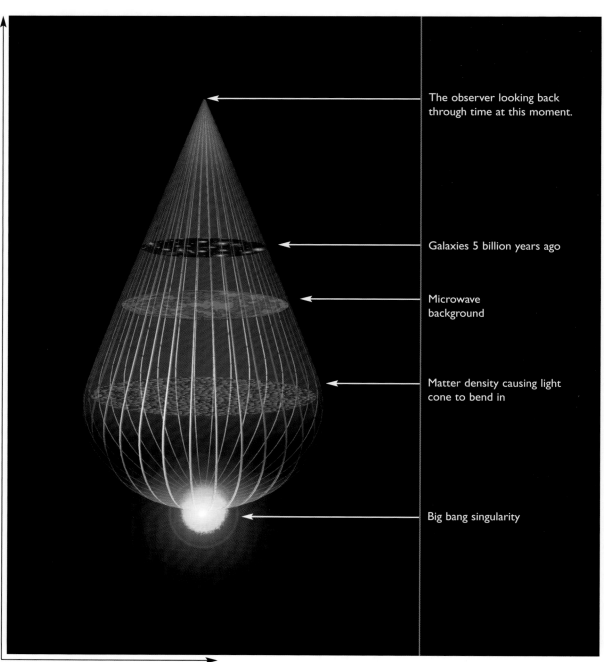

The observer looking back through time at this moment.

Galaxies 5 billion years ago

Microwave background

Matter density causing light cone to bend in

Big bang singularity

TIME

SPACE

As one goes back in time, the cross sections of our past light cone reach a maximum size and begin to get smaller again. Our past is pear-shaped (Fig. 2.8).

As one follows our past light cone back still further, the positive energy density of matter causes the light rays to bend toward each other more strongly. The cross section of the light cone will shrink to zero size in a finite time. This means that all the matter inside our past light cone is trapped in a region whose boundary shrinks to zero. It is therefore not very surprising that Penrose and I could prove that in the mathematical model of general relativity, time must have a beginning in what is called the big bang. Similar arguments show that time would have an end, when stars or galaxies collapse under their own gravity to form black holes. We had sidestepped Kant's antinomy of pure reason by dropping his implicit assumption that time had a meaning independent of the universe. Our paper, proving time had a beginning, won the second prize in the competition sponsored by the Gravity Research Foundation in 1968, and Roger and I shared the princely sum of $300. I don't think the other prize essays that year have shown much enduring value.

There were various reactions to our work. It upset many physicists, but it delighted those religious leaders who believed in an act of creation, for here was scientific proof. Meanwhile, Lifshitz and Khalatnikov were in an awkward position. They couldn't argue with the mathematical theorems that we had proved, but under the Soviet system they couldn't admit they had been wrong and Western science had been right. However, they saved the situation by finding a more general family of solutions with a singularity, which weren't special in the way their previous solutions had been. This enabled them to claim singularities, and the beginning or end of time, as a Soviet discovery.

(FIG. 2.8) TIME IS PEAR-SHAPED

If one follows our past light cone back in time, it will be bent back by the matter in the early universe. The whole universe we observe is contained within a region whose boundary shrinks to zero at the big bang. This would be a singularity, a place where the density of matter would be infinite and classical general relativity would break down.

THE UNCERTAINTY PRINCIPLE

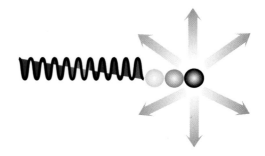

Low-frequency wavelengths disturb the velocity of the particle less.

High-frequency wavelengths disturb the velocity of the particle more.

The longer the wavelength used to observe a particle, the greater the uncertainty of its position.

The shorter the wavelength used to observe a particle, the greater the certainty of its position.

An important step in the discovery of quantum theory was Max Planck's suggestion in 1900 that light always comes in little packets he called quanta. But while Planck's quantum hypothesis clearly explained observations of the rate of radiation from hot bodies, the full extent of its implications wasn't realized until the mid-1920s, when the Germanphysicist Werner Heisenberg formulated his famous uncertainty principle. He noted that Planck's hypothesis implies that the more accurately one tries to measure the position of a particle, the less accurately one can measure its speed, and vice versa.

More precisely, he showed that the uncertainty in the position of a particle times the uncertainty in its momentum must always be larger than Planck's constant, which is a quantity that is closely related to the energy content of one quantum of light.

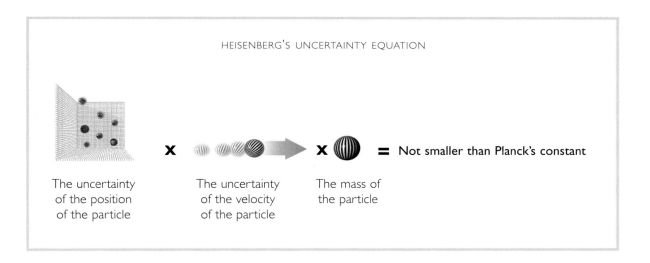

HEISENBERG'S UNCERTAINTY EQUATION

X → X = Not smaller than Planck's constant

The uncertainty of the position of the particle

The uncertainty of the velocity of the particle

The mass of the particle

Most physicists still instinctively disliked the idea of time having a beginning or end. They therefore pointed out that the mathematical model might not be expected to be a good description of spacetime near a singularity. The reason is that general relativity, which describes the gravitational force, is a classical theory, as noted in Chapter 1, and does not incorporate the uncertainty of quantum theory that governs all other forces we know. This inconsistency does not matter in most of the universe most of the time, because the scale on which spacetime is curved is very large and the scale on which quantum effects are important is very small. But near a singularity, the two scales would be comparable, and quantum gravitational effects would be important. So what the singularity theorems of Penrose and myself really established is that our classical region of spacetime is bounded to the past, and possibly to the future, by regions in which quantum gravity is important. To understand the origin and fate of the universe, we need a quantum theory of gravity, and this will be the subject of most of this book.

Quantum theories of systems such as atoms, with a finite number of particles, were formulated in the 1920s, by Heisenberg, Schrödinger, and Dirac. (Dirac was another previous holder of my chair in Cambridge, but it still wasn't motorized.) However, people encountered difficulties when they tried to extend quantum ideas to the Maxwell field, which describes electricity, magnetism, and light.

THE MAXWELL FIELD

In 1865 the British physicist James Clerk Maxwell combined all the known laws of electricity and magnetism. Maxwell's theory rests on the existence of "fields" that transmit actions from one place to another. He recognized that the fields that transmit electric and magnetic disturbances are dynamical entities: they can oscillate and move through space.

Maxwell's synthesis of electromagnetism can be condensed into two equations that prescribe the dynamics of these fields. He himself derived the first great conclusion from these equations: that electromagnetic waves of all frequencies travel through space at the same fixed speed—the speed of light.

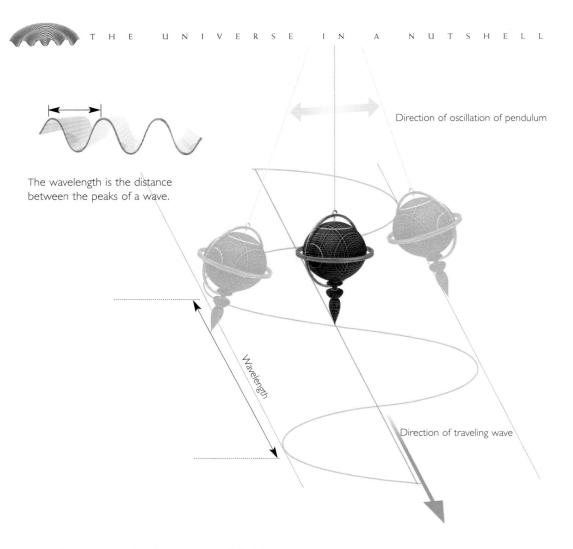

Direction of oscillation of pendulum

The wavelength is the distance between the peaks of a wave.

Wavelength

Direction of traveling wave

One can think of the Maxwell field as being made up of waves of different wavelengths (the distance between one wave crest and the next). In a wave, the field will swing from one value to another like a pendulum (Fig. 2.9).

According to quantum theory, the ground state, or lowest energy state, of a pendulum is not just sitting at the lowest energy point, pointing straight down. That would have both a definite position and a definite velocity, zero. This would be a violation of the uncertainty principle, which forbids the precise measurement of both position and velocity at the same time. The uncertainty in the position multiplied by the uncertainty in the momentum must be greater than a certain quantity, known as Planck's constant—a number that is too long to keep writing down, so we use a symbol for it: ħ

(FIG. 2.9)

TRAVELING WAVE WITH OSCILLATING PENDULUM

Electromagnetic radiation travels through space as a wave, with its electric and magnetic fields oscillating, like a pendulum, in directions transverse to the wave's direction of motion. The radiation can be made up of fields of different wavelengths.

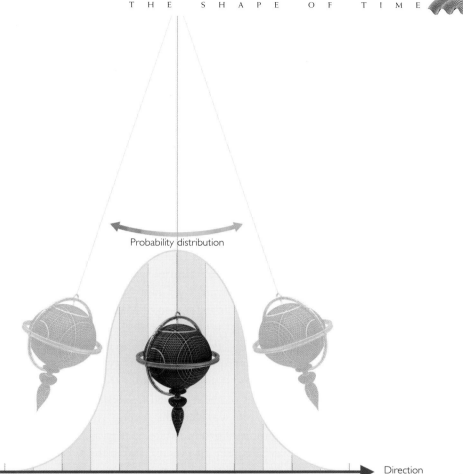

Probability distribution

Direction

So the ground state, or lowest energy state, of a pendulum does not have zero energy, as one might expect. Instead, even in its ground state a pendulum or any oscillating system must have a certain minimum amount of what are called zero point fluctuations. These mean that the pendulum won't necessarily be pointing straight down but will also have a probability of being found at a small angle to the vertical (Fig. 2.10). Similarly, even in the vacuum or lowest energy state, the waves in the Maxwell field won't be exactly zero but can have small sizes. The higher the frequency (the number of swings per minute) of the pendulum or wave, the higher the energy of the ground state.

Calculations of the ground state fluctuations in the Maxwell and electron fields made the apparent mass and charge of the electron infinite, which is not what observations show. However, in the

(FIG. 2.10)
PENDULUM WITH PROBABLITY DISTRIBUTION

According to the Heisenberg principle it is impossible for a pendulum to absolutely point straight down, with zero velocity. Instead quantum theory predicts that, even in its lowest energy state, the pendulum must have a minimum amount of fluctuations.

This means that the pendulum's position will be given by a probability distribution. In its ground state, the most likely position is pointing straight down, but it has also a probability of being found at a small angle to the vertical.

1940s the physicists Richard Feynman, Julian Schwinger, and Shin'ichiro Tomonaga developed a consistent way of removing or "subtracting out" these infinities and dealing only with the finite observed values of the mass and charge. Nevertheless, the ground state fluctuations still caused small effects that could be measured and that agreed well with experiment. Similar subtraction schemes for removing infinities worked for the Yang-Mills field in the theory put forward by Chen Ning Yang and Robert Mills. Yang-Mills theory is an extension of Maxwell theory that describes interactions in two other forces called the weak and strong nuclear forces. However, ground state fluctuations have a much more serious effect in a quantum theory of gravity. Again, each wavelength would have a ground state energy. Since there is no limit to how short the wavelengths of the Maxwell field can be, there are an infinite number of different wavelengths in any region of spacetime and an infinite amount of ground state energy. Because energy density is, like matter, a source of gravity, this infinite energy density ought to mean there is enough gravitational attraction in the universe to curl spacetime into a single point, which obviously hasn't happened.

One might hope to solve the problem of this seeming contradiction between observation and theory by saying that the ground state fluctuations have no gravitational effect, but this would not work. One can detect the energy of ground state fluctuations by the Casimir effect. If you place a pair of metal plates parallel to each other and close together, the effect of the plates is to reduce slightly the number of wavelengths that fit between the plates relative to the number outside. This means that the energy density of ground state fluctuations between the plates, although still infinite, is less than the energy density outside by a finite amount (Fig. 2.11). This difference in energy density gives rise to a force pulling the plates together, and this force has been observed experimentally. Forces are a source of gravity in general relativity, just as matter is, so it would not be consistent to ignore the gravitational effect of this energy difference.

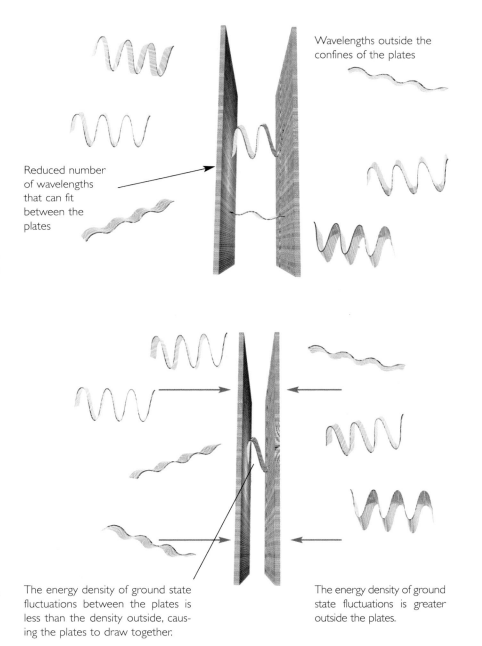

Reduced number
of wavelengths
that can fit
between the
plates

Wavelengths outside the
confines of the plates

(FIG. 2.11)
THE CASIMIR EFFECT

The existence of ground
state fluctuations has been
confirmed experimentally
by the Casimir effect, a slight
force between parallel
metal plates.

The energy density of ground state
fluctuations between the plates is
less than the density outside, causing the plates to draw together.

The energy density of ground
state fluctuations is greater
outside the plates.

47

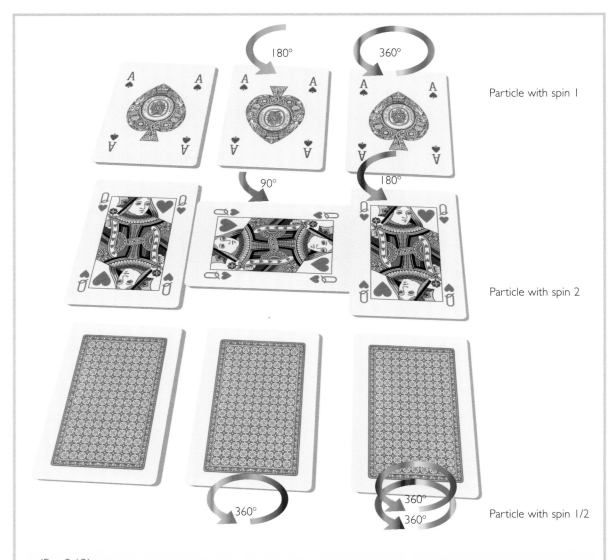

180°

360°

Particle with spin 1

90°

180°

Particle with spin 2

360°

360°

360°

Particle with spin 1/2

(FIG. 2.12) SPIN

All particles have a property called spin, having to do with what the particle looks like from different directions. One can illustrate this with a pack of playing cards. Consider first the ace of spades. This looks the same only if you turn it through a complete revolution, or 360 degrees. It is therefore said to have spin 1.

On the other hand, the queen of hearts has two heads. It is therefore the same under only half a revolution, 180

degrees. It is said to have spin 2. Similarly, one could imagine objects with spin 3 or higher that would look the same under smaller fractions of a revolution.

The higher the spin, the smaller the fraction of a complete revolution necessary to have the particle look the same. But the remarkable fact is that there are particles that look the same only if you turn them through two complete revolutions. Such particles are said to have spin 1/2.

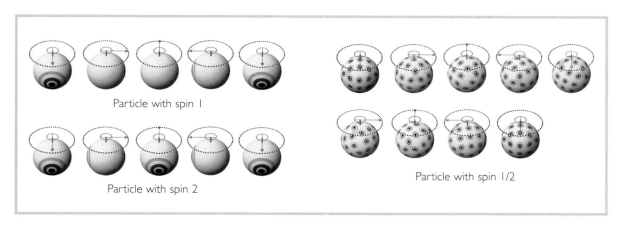

Particle with spin 1

Particle with spin 2

Particle with spin 1/2

Another possible solution to the problem might be to suppose there was a cosmological constant such as Einstein introduced in an attempt to have a static model of the universe. If this constant had an infinite negative value, it could exactly cancel the infinite positive value of the ground state energies in free space, but this cosmological constant seems very ad hoc, and it would have to be tuned to extraordinary accuracy.

Fortunately, a totally new kind of symmetry was discovered in the 1970s that provides a natural physical mechanism to cancel the infinities arising from ground state fluctuations. Supersymmetry is a feature of our modern mathematical models that can be described in various ways. One way is to say that spacetime has extra dimensions besides the dimensions we experience. These are called Grassmann dimensions, because they are measured in numbers known as Grassmann variables rather than in ordinary real numbers. Ordinary numbers commute; that is, it does not matter in which order you multiply them: 6 times 4 is the same as 4 times 6. But Grassmann variables *anti*commute: x times y is the same as −y times x.

Supersymmetry was first considered for removing infinities in matter fields and Yang-Mills fields in a spacetime where both the ordinary number dimensions and the Grassmann dimensions were flat, not curved. But it was natural to extend it to ordinary numbers and Grassmann dimensions that were curved. This led to a number of theories called supergravity, with different amounts of supersymmetry. One consequence of supersymmetry is that every field or particle should have a "superpartner" with a spin that is either 1/2 greater than its own or 1/2 less (Fig 2.12).

ORDINARY NUMBERS

$A \times B = B \times A$

GRASSMANN NUMBERS

$A \times B = -B \times A$

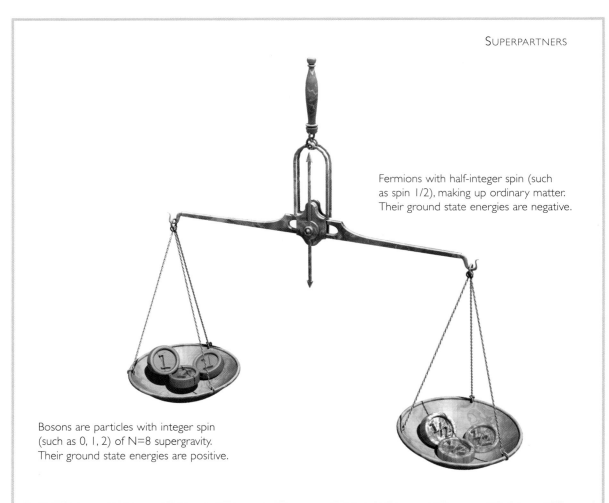

Fermions with half-integer spin (such as spin 1/2), making up ordinary matter. Their ground state energies are negative.

Bosons are particles with integer spin (such as 0, 1, 2) of N=8 supergravity. Their ground state energies are positive.

(FIG. 2.13)

All known particles in the universe belong to one of two groups, fermions or bosons. Fermions are particles with half-integer spin (such as spin 1/2), and they make up ordinary matter. Their ground state energies are negative.

Bosons are particles with integer spin (such as 0, 1, 2), and these give rise to forces between the fermions, such as the gravitational force and light. Their ground state energies are positive. Supergravity theory supposes that every fermion and every boson has a superpartner with a spin that is either 1/2 greater than its own or 1/2 less.

For example, a photon (which is a boson) has a spin of 1. Its ground state energy is positive. The photon's superpartner, the photino, has a spin of 1/2, making it a fermion. Hence, its ground state energy is negative.

In this supergravity scheme we end up with equal numbers of bosons and fermions. With the ground state energies of the bosons weighing in on the positive side and the fermions weighing in on the negative side, the ground state energies cancel one another out, eliminating the biggest infinities.

MODELS OF PARTICLE BEHAVIOR

1 If point particles actually existed as discrete elements like pool balls, then when two collided their path would be deflected into two new trajectories.

2 This is what appears to happen when two particles interact, although the effect is far more dramatic.

3 Quantum field theory shows two particles, like an electron and its antiparticle, a positron, colliding. In doing so they briefly annihilate one another in a frantic burst of energy, creating a photon. This then releases its energy, producing another electron-positron pair. This still appears as if they are just deflected into new trajectories.

4 If particles are not zero points but one-dimensional strings in which the oscillating loops vibrate as an electron and positron. Then, when they collide and annihilate one another, they create a new string with a different vibrational pattern. Releasing energy, it divides into two strings continuing along new trajectories.

5 If those original strings are viewed not as discrete moments but as an uninterrupted history in time, then the resulting strings are seen as a string world sheet.

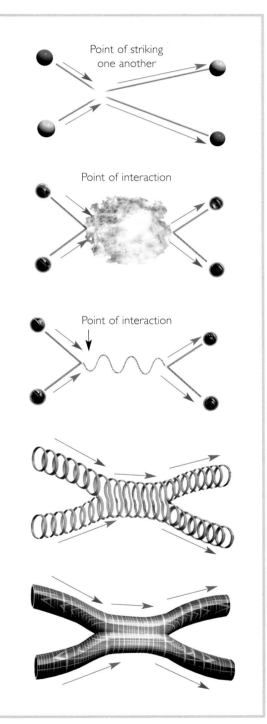

Point of striking
one another

Point of interaction

Point of interaction

(FIG. 2.14, opposite)
STRING OSCILLATIONS

In string theory the basic objects are not particles, which occupy a single point in space, but one-dimensional strings. These strings may have ends or they may join up with themselves in closed loops.

Just like the strings on a violin, the strings in string theory support certain vibrational patterns, or resonant frequencies, whose wavelengths fit precisely between the two ends.

But while the different resonant frequencies of a violin's strings give rise to different musical notes, the different oscillations of a string give rise to different masses and force charges, which are interpreted as fundamental particles. Roughly speaking, the shorter the wavelength of the oscillation on the string, the greater the mass of the particle.

The ground state energies of bosons, fields whose spin is a whole number (0, 1, 2 , etc.), are positive. On the other hand, the ground state energies of fermions, fields whose spin is a half number (1/2, 3/2 , etc.), are negative. Because there are equal numbers of bosons and fermions, the biggest infinities cancel in supergravity theories (see Fig 2.13, page 50).

There remained the possibility that there might be smaller but still infinite quantities left over. No one had the patience needed to calculate whether these theories were actually completely finite. It was reckoned it would take a good student two hundred years, and how would you know he hadn't made a mistake on the second page? Still, up to 1985, most people believed that most supersymmetric supergravity theories would be free of infinities.

Then suddenly the fashion changed. People declared there was no reason not to expect infinities in supergravity theories, and this was taken to mean they were fatally flawed as theories. Instead, it was claimed that a theory named supersymmetric string theory was the only way to combine gravity with quantum theory. Strings, like their namesakes in everyday experience, are one-dimensional extended objects. They have only length. Strings in string theory move through a background spacetime. Ripples on the string are interpreted as particles (Fig. 2.14).

If the strings have Grassmann dimensions as well as their ordinary number dimensions, the ripples will correspond to bosons and fermions. In this case, the positive and negative ground state energies will cancel so exactly that there will be no infinities even of the smaller sort. Superstrings, it was claimed, were the TOE, the Theory of Everything.

Historians of science in the future will find it interesting to chart the changing tide of opinion among theoretical physicists. For a few years, strings reigned supreme and supergravity was dismissed as just an approximate theory, valid at low energy. The qualification "low energy" was considered particularly damning, even though in this context low energies meant particles with energies of

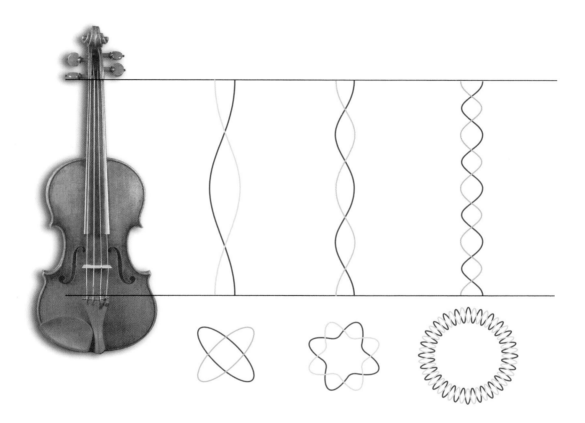

less than a billion billion times those of particles in a TNT explosion. If supergravity was only a low energy approximation, it could not claim to be the fundamental theory of the universe. Instead, the underlying theory was supposed to be one of five possible superstring theories. But which of the five string theories described our universe? And how could string theory be formulated, beyond the approximation in which strings were pictured as surfaces with one space dimension and one time dimension moving through a flat background spacetime? Wouldn't the strings curve the background spacetime?

In the years after 1985, it gradually became apparent that string theory wasn't the complete picture. To start with, it was realized that strings are just one member of a wide class of objects that can be extended in more than one dimension. Paul Townsend, who, like me, is a member of the Department of Applied Mathematics and Theoretical Physics at Cambridge, and who did much of the fundamental work on these objects, gave them the name "p-branes." A p-brane has length in p directions. Thus a p=1 brane is a string, a p=2 brane is a surface or membrane, and so on (Fig. 2.15). There seems no reason to favor the p=1 string case over other possible values of p. Instead, we should adopt the principle of p-brane democracy: all p-branes are created equal.

All the p-branes could be found as solutions of the equations of supergravity theories in 10 or 11 dimensions. While 10 or 11 dimensions doesn't sound much like the spacetime we experience, the idea was that the other 6 or 7 dimensions are curled up so small that we don't notice them; we are only aware of the remaining 4 large and nearly flat dimensions.

I must say that personally, I have been reluctant to believe in extra dimensions. But as I am a positivist, the question "Do extra dimensions really exist?" has no meaning. All one can ask is whether mathematical models with extra dimensions provide a good description of the universe. We do not yet have any observations that require extra dimensions for their explanation. However, there is a possibility we may observe them in the Large Hadron Collider

(FIG. 2.15) P-BRANES

P-branes are objects that are extended in p dimensions. Special cases are strings, which are p=1, and membranes, which are p=2, but higher values of p are possible in ten- or eleven-dimensional spacetime. Often, some or all of the p-dimensions are curled up like a torus.

*We hold these truths
to be self-evident:
All p-branes
are created equal!*

Paul Townsend, the egghead of p-branes

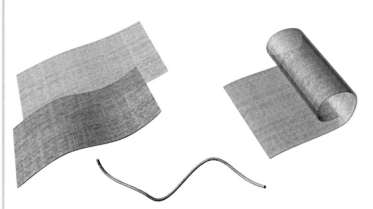

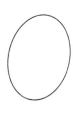

The spatial fabric of our universe may have both extended and curled-up dimensions. The membranes can be seen better if they are curled up.

A 1-brane or string curled up

A 2-brane sheet curled up into a torus

(FIG. 2.16) A UNIFIED FRAMEWORK?

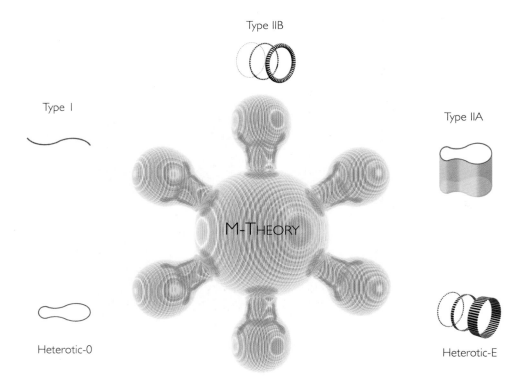

Type IIB

Type I

Type IIA

M-Theory

Heterotic-0

Heterotic-E

11-dimensional supergravity

There is a web of relationships, so-called dualities, that connect all five string theories as well as eleven-dimensional supergravity. The dualities suggest that the different string theories are just different expressions of the same underlying theory, which has been named M-theory.

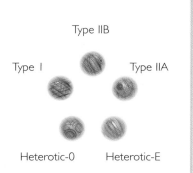

Type IIB

Type I

Type IIA

Heterotic-0   Heterotic-E

Prior to the mid-nineties it appeared that there were five distinct string theories, each separate and unconnected.

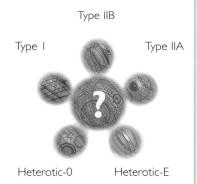

Type IIB

Type I

Type IIA

Heterotic-0   Heterotic-E

M-theory unites the five string theories within a single theoretical framework, but many of its properties have yet to be understood.

in Geneva. But what has convinced many people, including myself, that one should take models with extra dimensions seriously is that there is a web of unexpected relationships, called dualities, between the models. These dualities show that the models are all essentially equivalent; that is, they are just different aspects of the same underlying theory, which has been given the name M-theory. Not to take this web of dualities as a sign we are on the right track would be a bit like believing that God put fossils into the rocks in order to mislead Darwin about the evolution of life.

These dualities show that the five superstring theories all describe the same physics and that they are also physically equivalent to supergravity (Fig. 2.16). One cannot say that superstrings are more fundamental than supergravity, or vice versa. Rather, they are different expressions of the same underlying theory, each useful for calculations in different kinds of situations. Because string theories don't have any infinities, they are good for calculating what happens when a few high energy particles collide and scatter off each other. However, they are not of much use for describing how the energy of a very large number of particles curves the universe or forms a bound state, like a black hole. For these situations, one needs supergravity, which is basically Einstein's theory of curved spacetime with some extra kinds of matter. It is this picture that I shall mainly use in what follows.

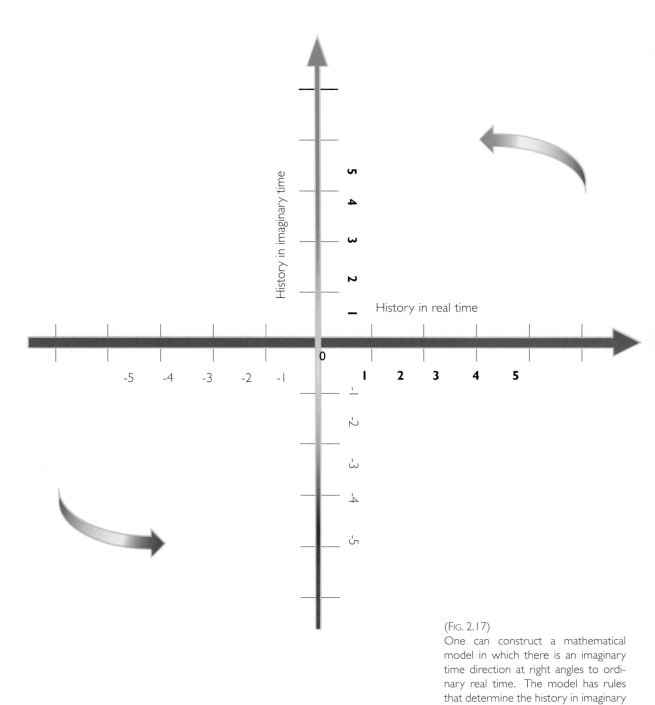

History in imaginary time

History in real time

5
4
3
2
1

-5   -4   -3   -2   -1   0   1   2   3   4   5

-1
-2
-3
-4
-5

(Fig. 2.17)
One can construct a mathematical model in which there is an imaginary time direction at right angles to ordinary real time. The model has rules that determine the history in imaginary time in terms of the history in real time, and vice versa.

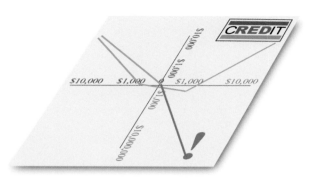

To describe how quantum theory shapes time and space, it is helpful to introduce the idea of imaginary time. Imaginary time sounds like something from science fiction, but it is a well-defined mathematical concept: time measured in what are called imaginary numbers. One can think of ordinary real numbers such as 1, 2, -3.5, and so on as corresponding to positions on a line stretching from left to right: zero in the middle, positive real numbers on the right, and negative real numbers on the left (Fig. 2.17).

Imaginary numbers can then be represented as corresponding to positions on a vertical line: zero is again in the middle, positive imaginary numbers plotted upward, and negative imaginary numbers plotted downward. Thus imaginary numbers can be thought of as a new kind of number at right angles to ordinary real numbers. Because they are a mathematical construct, they don't need a physical realization; one can't have an imaginary number of oranges or an imaginary credit card bill (Fig. 2.18).

One might think this means that imaginary numbers are just a mathematical game having nothing to do with the real world. From the viewpoint of positivist philosophy, however, one cannot determine what is real. All one can do is find which mathematical models describe the universe we live in. It turns out that a mathematical model involving imaginary time predicts not only effects we have already observed but also effects we have not been able to measure yet nevertheless believe in for other reasons. So what is real and what is imaginary? Is the distinction just in our minds?

(FIG. 2.18)
Imaginary numbers are a mathematical construction. You can't have an imaginary number credit card bill.

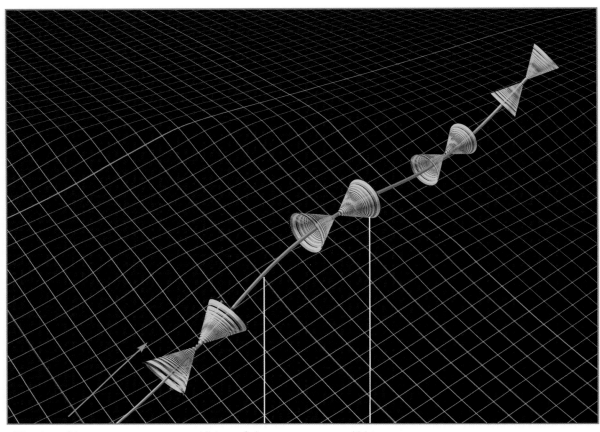

Direction of time   History of observer   Light cones

(FIG. 2.19)

In the real time spacetime of classical general relativity, time is distinguished from the space directions because it increases only along the history of an observer, unlike the space directions, which can increase or decrease along that history. The imaginary time direction of quantum theory, on the other hand, is like another space direction, so can increase or decrease.

Einstein's classical (i.e., nonquantum) general theory of relativity combined real time and the three dimensions of space into a four-dimensional spacetime. But the real time direction was distinguished from the three spatial directions; the world line or history of an observer always increased in the real time direction (that is, time always moved from past to future), but it could increase *or decrease* in any of the three spatial directions. In other words, one could reverse direction in space, but not in time (Fig. 2.19).

On the other hand, because imaginary time is at right angles to real time, it behaves like a fourth spatial direction. It can therefore

(FIG. 2.20) IMAGINARY TIME

In an imaginary spacetime that is a sphere, the imaginary time direction could represent the distance from the South Pole. As one moves north, the circles of latitude at constant distances from the South Pole become bigger, corresponding to the universe expanding with imaginary time. The universe would reach maximum size at the equator and then contract again with increasing imaginary time to a single point at the North Pole. Even though the universe would have zero size at the poles, these points would not be singularities, just as the North and South Poles on the Earth's surface are perfectly regular points. This suggests that the origin of the universe in imaginary time can be a regular point in spacetime.

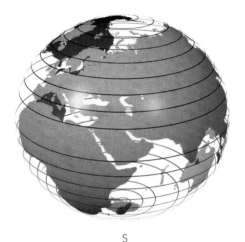

S

Imaginary time as degrees of latitude

N

(FIG. 2.21)

Instead of degrees of latitude, the imaginary time direction in a spacetime that is a sphere could also correspond to degrees of longitude. Because all the lines of longitude meet at the North and South Poles, time is standing still at the poles; an increase of imaginary time leaves one on the same spot, just as going west on the North Pole of the Earth still leaves one on the North Pole.

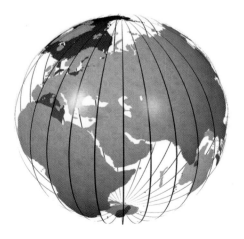

Imaginary time as degrees of longitude which meet at the North and South Poles

61

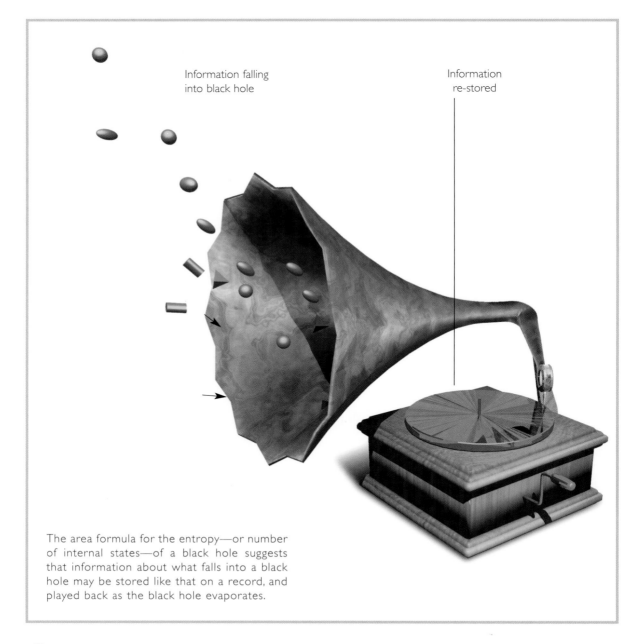

Information falling
into black hole

Information
re-stored

The area formula for the entropy—or number
of internal states—of a black hole suggests
that information about what falls into a black
hole may be stored like that on a record, and
played back as the black hole evaporates.

have a much richer range of possibilities than the railroad track of ordinary real time, which can only have a beginning or an end or go around in circles. It is in this imaginary sense that time has a shape.

To see some of the possibilities, consider an imaginary time spacetime that is a sphere, like the surface of the Earth. Suppose that imaginary time was degrees of latitude (Fig. 2.20, see page 61). Then the history of the universe in imaginary time would begin at the South Pole. It would make no sense to ask, "What happened before the beginning?" Such times are simply not defined, any more than there are points south of the South Pole. The South Pole is a perfectly regular point of the Earth's surface, and the same laws hold there as at other points. This suggests that the beginning of the universe in imaginary time can be a regular point of spacetime, and that the same laws can hold at the beginning as in the rest of the universe. (The quantum origin and evolution of the universe will be discussed in the next chapter.)

Another possible behavior is illustrated by taking imaginary time to be degrees of longitude on the Earth. All the lines of longitude meet at the North and South Poles (Fig. 2.21, see page 61) Thus time stands still there, in the sense that an increase of imaginary time, or of degrees of longitude, leaves one in the same spot. This is very similar to the way that ordinary time appears to stand still on the horizon of a black hole. We have come to recognize that this standing still of real and imaginary time (either both stand still or neither does) means that the spacetime has a temperature, as I discovered for black holes. Not only does a black hole have a temperature, it also behaves as if it has a quantity called entropy. The entropy is a measure of the number of internal states (ways it could be configured on the inside) that the black hole could have without looking any different to an outside observer, who can only observe its mass, rotation, and charge. This black hole entropy is given by a very simple formula I discovered in 1974. It equals the area of the horizon of the black hole: there is one bit of information about the internal state of the black hole for each fundamental unit of area of

$$S = \frac{A k c^3}{4 \hbar G}$$

THE BLACK HOLE ENTROPY FORMULA

A the area of the event horizon of the black hole

$\hbar$ Planck's constant

k Boltzmann's constant

G Newton's gravitational constant

c Speed of light

S Entropy

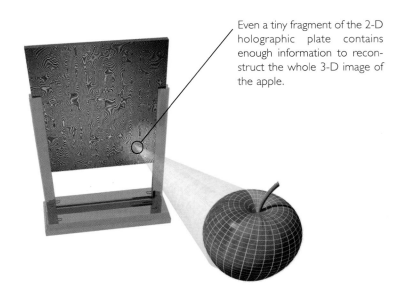

Even a tiny fragment of the 2-D holographic plate contains enough information to reconstruct the whole 3-D image of the apple.

THE HOLOGRAPHIC PRINCIPLE

The realization that the surface area of the horizon surrounding a black hole measures the black hole's entropy has led people to advocate that the maximum entropy of any closed region of space can never exceed a quarter of the area of the circumscribing surface. Since entropy is nothing more than a measure of the total information contained in a system, this suggests that the information associated with all phenomena in the three-dimensional world can be stored on its two-dimensional boundary, like a holographic image. In a certain sense the world would be two-dimensional.

the horizon. This shows that there is a deep connection between quantum gravity and thermodynamics, the science of heat (which includes the study of entropy). It also suggests that quantum gravity may exhibit what is called holography (Fig. 2.22).

Information about the quantum states in a region of spacetime may be somehow coded on the boundary of the region, which has two dimensions less. This is like the way that a hologram carries a three-dimensional image on a two-dimensional surface. If quantum gravity incorporates the holographic principle, it may mean that we can keep track of what is inside black holes. This is essential if we are to be able to predict the radiation that comes out of black holes. If we can't do that, we won't be able to predict the future as fully as we thought. This is discussed in Chapter 4. Holography is discussed again in Chapter 7. It seems we may live on a 3-brane—a four-dimensional (three space plus one time) surface that is the boundary of a five-dimensional region, with the remaining dimensions curled up very small. The state of the world on a brane encodes what is happening in the five-dimensional region.

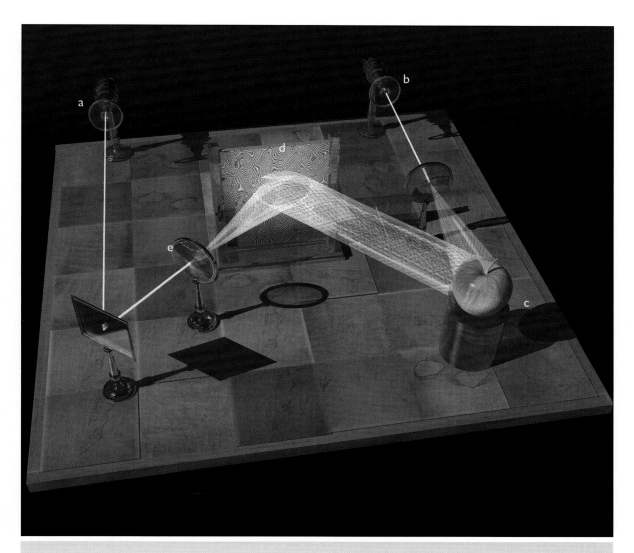

(FIG. 2.22) Holography is essentially a phenomenon of interference of wave patterns. Holograms are created when the light from a single laser is split into two separate beams (a) and (b). One (b) bounces off the object (c) onto a photo-sensitized plate (d). The other (a) passes through a lens (e) and collides with the reflected light of (b), creating an interference pattern on the plate.

When a laser is shone through the developed plate a fully *three-dimensional* image of the original object appears. An observer can move around this holographic image, being able to see all the hidden faces that a normal photo could not show.

The two-dimensional surface of the plate on the left, unlike a normal photo, has the remarkable property that any tiny fragment of its surface contains all the information needed to reconstruct the whole image.

CHAPTER 3

THE UNIVERSE IN A NUTSHELL

The universe has multiple histories, each of which is determined by a tiny nut.

*I could be bounded in a nutshell
and count myself a king of infinite space...*

—Shakespeare,
Hamlet, Act 2, Scene 2

Hamlet may have meant that although we human beings are very limited physically, our minds are free to explore the whole universe, and to go boldly where even *Star Trek* fears to tread—bad dreams permitting.

Is the universe actually infinite or just very large? And is it everlasting or just long-lived? How could our finite minds comprehend an infinite universe? Isn't it presumptuous of us even to make the attempt? Do we risk the fate of Prometheus, who in classical mythology stole fire from Zeus for human beings to use, and was punished for his temerity by being chained to a rock where an eagle picked at his liver?

Despite this cautionary tale, I believe we can and should try to understand the universe. We have already made remarkable progress in understanding the cosmos, particularly in the last few years. We don't yet have a complete picture, but this may not be far off.

The most obvious thing about space is that it goes on and on and on. This has been confirmed by modern instruments such as the Hubble telescope, which allows us to probe deep into space. What we see are billions and billions of galaxies of various shapes and sizes (see page 70, Fig. 3.1). Each galaxy contains uncounted billions of stars, many of which have planets around them. We live on a planet orbiting a star in an outer arm of the spiral Milky Way

Above: *Prometheus. Etruscan vase painting, 6th century* B.C.

Left: *Hubble space telescope lens and mirrors being upgraded by a space shuttle mission. Australia can be seen below.*

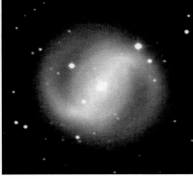

Spiral galaxy NGC 4414   *Spiral bar galaxy NGC 4314*   *Elliptical galaxy NGC 147*

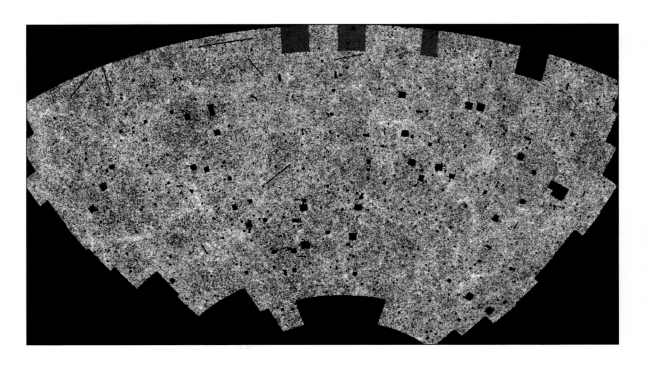

(FIG. 3.1) When we look deep into the universe, we see billions and billions of galaxies.
Galaxies can have various shapes and sizes; they can be either elliptical or spiral, like our own Milky Way.

galaxy. The dust in the spiral arms blocks our view of the universe in the plane of the galaxy, but we have a clear line of sight in cones of directions on each side of the plane, and we can plot the positions of distant galaxies (Fig. 3.2). We find that the galaxies are distributed roughly uniformly throughout space, with some local concentrations and voids. The density of galaxies appears to drop off at very large distances, but that seems to be because they are so far away and faint that we can't make them out. As far as we can tell, the universe goes on in space forever (see page 72, Fig. 3.3).

Although the universe seems to be much the same at each position in space, it is definitely changing in time. This was not realized until the early years of the twentieth century. Up to then, it was thought the universe was essentially constant in time. It might have existed for an infinite time, but that seemed to lead to absurd conclusions. If stars had been radiating for an infinite time, they would have heated up the universe to their temperature. Even

(FIG. 3.2)
Our planet Earth (**E**) orbits the Sun in the outer region of the spiral Milky Way galaxy. The stellar dust in the spiral arms blocks our view within the plane of the galaxy but we have a clear view on either side of that plane.

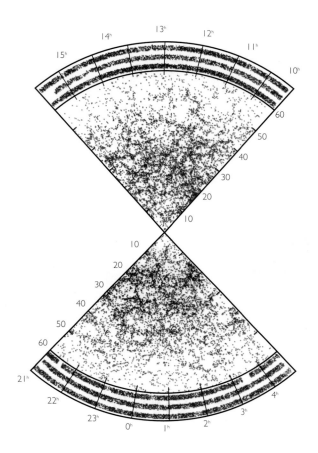

(FIG. 3.3)

Apart from some local concentrations, we find that galaxies are distributed roughly uniformly throughout space.

at night, the whole sky would be as bright as the sun, because every line of sight would end either on a star or on a cloud of dust that had been heated up until it was as hot as the stars (Fig. 3.4).

The observation that we have all made, that the sky at night is dark, is very important. It implies that the universe cannot have existed forever in the state we see today. Something must have happened in the past to make the stars light up a finite time ago, which means that the light from very distant stars has not had time to reach us yet. This would explain why the sky at night isn't glowing in every direction.

(FIG. 3.4)

If the universe was static and infinite in every direction, every line of sight would end in a star, which would make the night sky as bright as the sun.

If the stars had just been sitting there forever, why did they suddenly light up a few billion years ago? What was the clock that told them it was time to shine? As we've seen, this puzzled those philosophers, much like Immanuel Kant, who believed that the universe had existed forever. But for most people, it was consistent with the idea that the universe had been created, much as it is now, only a few thousand years ago.

However, discrepancies with this idea began to appear with the observations by Vesto Slipher and Edwin Hubble in the second decade of the twentieth century. In 1923, Hubble discovered that

THE DOPPLER EFFECT

The relationship between speed and wavelength, which is called the Doppler effect, is an everyday experience.

Listen to a plane that passes overhead; as it approaches, its engine sounds at a higher pitch, and when it passes and disappears, it sounds at a lower pitch.

The higher pitch corresponds to sound waves with a shorter wavelength (the distance between one wave crest and the next) and a higher frequency (the number of waves per second).

This is because, as the plane moves toward you, it will be nearer to you when it emits the next wave crest, lessening the distance between wave crests.

Similarly, as the plane moves away the wavelengths increase and the pitch you perceive is lower.

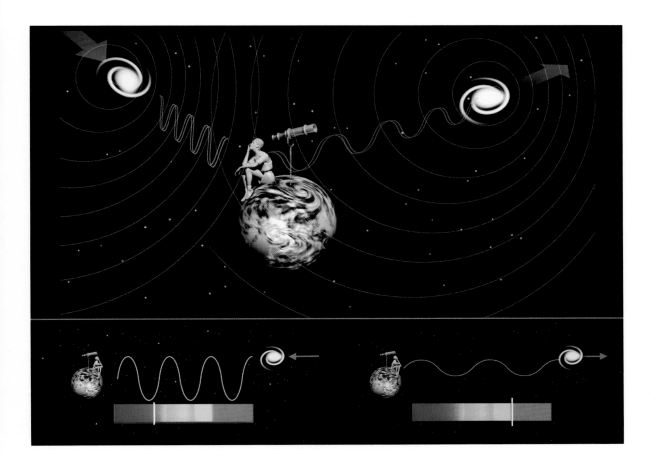

many faint patches of light, called nebulae, were in fact other galaxies, vast collections of stars like our sun but at a great distance. In order for them to appear so small and faint, the distances had to be so great that light from them would have taken millions or even billions of years to reach us. This indicated that the beginning of the universe couldn't have been just a few thousand years ago.

But the second thing Hubble discovered was even more remarkable. Astronomers had learned that by analyzing the light from other galaxies, it was possible to measure whether they are moving toward us or away from us (Fig. 3.5). To their great surprise, they had found that nearly all galaxies are moving away. Moreover, the farther they are from us, the faster they are moving away. It was Hubble who recognized the dramatic implications of this discovery:

(FIG. 3.5)
The Doppler effect is also true of light waves. If a galaxy were to remain at a constant distance from Earth, characteristic lines in the spectrum would appear in a normal or standard position. However, if the galaxy is moving away from us, the waves will appear elongated or stretched and the characteristic lines will be shifted toward the red (right). If the galaxy is moving toward us then the waves will appear to be compressed, and the lines will be blue-shifted (left).

Our galactic neighbor, Andromeda, measured by Hubble and Slipher.

CHRONOLOGY OF DISCOVERIES MADE BY SLIPHER AND HUBBLE, BETWEEN 1910 AND 1930.

1912 —Slipher measured light from four nebulae, finding three of them red-shifted but Andromeda blue-shifted. His interpretation was that Andromeda is moving toward us while the other nebulae move away from us.

1912-1914 —Slipher measured 12 more nebulae. All except one were red-shifted.

1914 —Slipher presented his findings to the American Astronomical Society. Hubble heard the presentation.

1918 —Hubble began to investigate the nebulae.

1923 —Hubble determined that the spiral nebulae (including Andromeda) are other galaxies.

1914-1925 —Slipher and others kept measuring Doppler shifts. The score in 1925 was 43 red shifts to 2 blue shifts.

1929 —Hubble and Milton Humason—after continuing to measure Doppler shifts and finding that on the large scale every galaxy appears to be receding from every other—announced their discovery that the universe is expanding.

on the large scale, every galaxy is moving away from every other galaxy. The universe is expanding (Fig. 3.6).

The discovery of the expansion of the universe was one of the great intellectual revolutions of the twentieth century. It came as a total surprise, and it completely changed the discussion of the origin of the universe. If the galaxies are moving apart, they must have been closer together in the past. From the present rate of expansion, we can estimate that they must have been very close together indeed ten to fifteen billion years ago. As described in the last chapter, Roger Penrose and I were able to show that Einstein's general theory of relativity implied that the universe and time itself must have had a beginning in a tremendous explosion. Here was the explanation of

Edwin Hubble at the 100-inch Mount Wilson telescope in 1930.

(FIG. 3.6) HUBBLE'S LAW

By analyzing the light from other galaxies, Edwin Hubble discovered in the 1920s that nearly all galaxies are moving away from us, at a velocity **V** that is proportional to their distance **R** from the Earth, so **V** = **H** × **R**.

This important observation, known as Hubble's law, established that the universe is expanding,

with the Hubble constant **H** setting the rate of expansion.

The graph below shows recent observations of the red-shift of galaxies, confirming Hubble's law to vast distances away from us.

The slight upward bend in the graph at large distances indicates that the expansion is speeding up, which may be caused by vacuum energy.

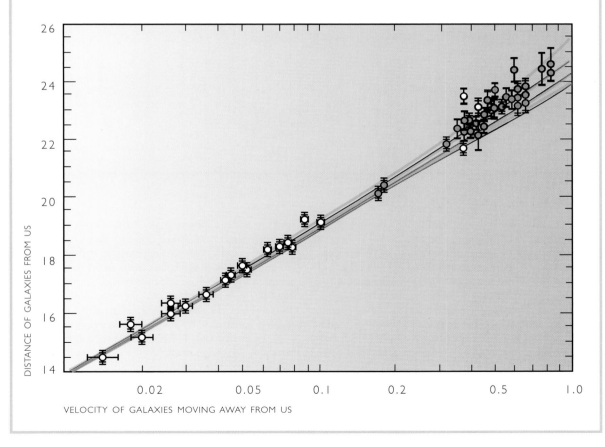

DISTANCE OF GALAXIES FROM US

VELOCITY OF GALAXIES MOVING AWAY FROM US

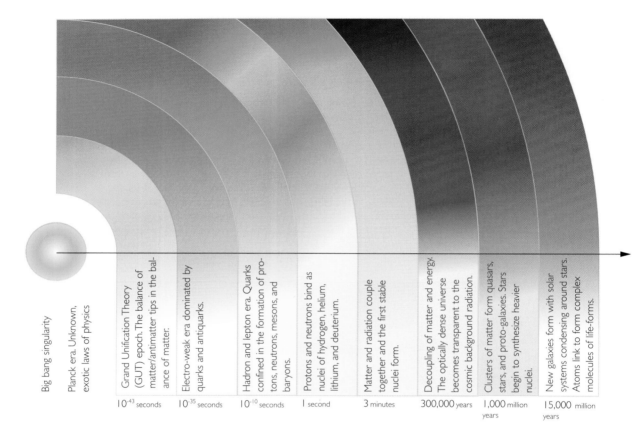

Big bang singularity

Planck era. Unknown, exotic laws of physics

Grand Unification Theory (GUT) epoch. The balance of matter/antimatter tips in the balance of matter.

Electro-weak era dominated by quarks and antiquarks.

Hadron and lepton era. Quarks confined in the formation of protons, neutrons, mesons, and baryons.

Protons and neutrons bind as nuclei of hydrogen, helium, lithium, and deuterium.

Matter and radiation couple together and the first stable nuclei form.

Decoupling of matter and energy. The optically dense universe becomes transparent to the cosmic background radiation.

Clusters of matter form quasars, stars, and proto-galaxies. Stars begin to synthesize heavier nuclei.

New galaxies form with solar systems condensing around stars. Atoms link to form complex molecules of life-forms.

10^{-43} seconds 10^{-35} seconds 10^{-10} seconds   I second   3 minutes   300,000 years   1,000 million years   15,000 million years

THE HOT BIG BANG

If general relativity were correct, the universe started with an infinite temperature and density at the big bang singularity. As the universe expanded, the temperature of the radiation decreased. At about a hundredth of a second after the big bang, the temperature would have been 100 billion degrees, and the universe would have contained mostly photons, electrons, and neutrinos (extremely light particles), and their antiparticles, together with some protons and neutrons. For the next three minutes, as the universe cooled to about one billion degrees, protons and neutrons would have begun combining to form the nuclei of helium, hydrogen, and other light elements.

Hundreds of thousands of years later, when the tem-perature had dropped to a few thousand degrees, the electrons would have slowed down to the point where the light nuclei could capture them to form atoms. However, the heavier elements of which we are made, such as carbon and oxygen, would not form until billion of years later from the burning of helium in the center of stars.

This picture of a dense, hot early stage of the universe was first put forward by the scientist George Gamow in 1948, in a paper he wrote with Ralph Alpher, which made the remarkable prediction that radiation from this very hot early stage should still be around today. Their prediction was confirmed in 1965, when the physicists Arno Penzias and Robert Wilson observed the cosmic microwave background radiation.

why the sky at night is dark: no star could have been shining longer than ten to fifteen billion years, the time since the big bang.

We are used to the idea that events are caused by earlier events, which in turn are caused by still earlier events. There is a chain of causality stretching back into the past. But suppose this chain has a beginning. Suppose there was a first event. What caused *it?* This was not a question that many scientists wanted to address. They tried to avoid it, either by claiming, like the Russians, that the universe didn't have a beginning or by maintaining that the origin of the universe did not lie within the realm of science but belonged to metaphysics or religion. In my opinion, this is not a position any true scientist should take. If the laws of science are suspended at the beginning of the universe, might not they fail at other times also? A law is not a law if it only holds sometimes. *We must try to understand the beginning of the universe on the basis of science. It may be a task beyond our powers, but we should at least make the attempt.*

While the theorems that Penrose and I proved showed that the universe must have had a beginning, they didn't give much information about the nature of that beginning. They indicated that the universe began in a big bang, a point where the whole universe, and everything in it, was scrunched up into a single point of infinite density. At this point, Einstein's general theory of relativity would have broken down, so it cannot be used to predict in what manner the universe began. One is left with the origin of the universe apparently being beyond the scope of science.

This was not a conclusion that scientists should be happy with. As Chapters 1 and 2 point out, the reason general relativity broke down near the big bang is that it did not incorporate the uncertainty principle, the random element of quantum theory that Einstein had objected to on the grounds that God does not play dice. However, all the evidence is that God is quite a gambler. One can think of the universe as being like a giant casino, with dice being rolled or wheels being spun on every occasion (Fig. 3.7). You might think that operating a casino is a very chancy business, because you risk losing money each time dice are thrown or the wheel is spun. But over a large number of bets, the gains and losses average out to a result that *can* be predicted, even though the result of any particular bet cannot be predicted (Fig. 3.8). The casino operators make sure the odds average out in their favor. That is why casino operators are so rich. The only chance you have of winning against them is to stake all your money on a few rolls of the dice or spins of the wheel.

It is the same with the universe. When the universe is big, as it is today, there are a very large number of rolls of the dice, and the results average out to something one can predict. That is why classical laws work for large systems. But when the universe is very small, as it was near in time to the big bang, there are only a small number of rolls of the dice, and the uncertainty principle is very important.

Because the universe keeps on rolling the dice to see what happens next, it doesn't have just a single history, as one might have thought. Instead, the universe must have every possible history, each with its own probability. There must be a history of the universe in which Belize won every gold medal at the Olympic Games, though maybe the probability is low.

This idea that the universe has multiple histories may sound like science fiction, but it is now accepted as science fact. It was formulated by Richard Feynman, who was both a great physicist and quite a character.

We are now working to combine Einstein's general theory of relativity and Feynman's idea of multiple histories into a complete unified theory that will describe everything that happens in the universe. This unified theory will enable us to calculate how the universe will develop if we know how the histories started. But the unified theory will not in itself tell us how the universe began or what its initial state was. For that, we need what are called boundary conditions, rules that tell us what happens on the frontiers of the universe, the edges of space and time.

If the frontier of the universe was just at a normal point of space and time, we could go past it and claim the territory beyond as part of the universe. On the other hand, if the boundary of the

(FIG. 3.7, above, and FIG. 3.8, opposite) If a gambler bets on red for a large number of spins of the wheel, one can fairly accurately predict his return because the results of the single spins average out.

On the other hand, it is impossible to predict the outcome of any particular bet.

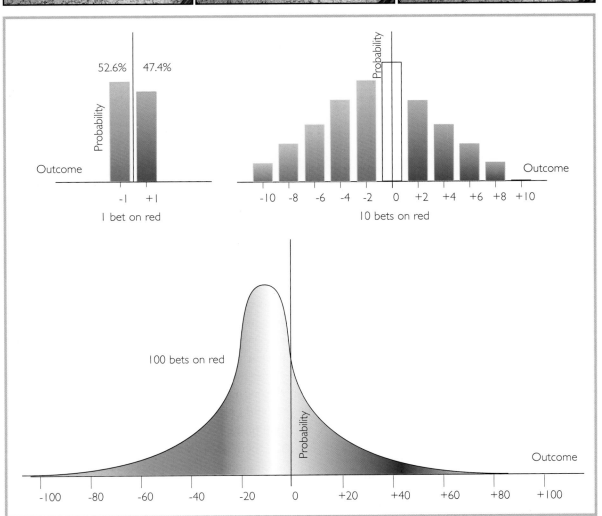

52.6%   47.4%

Probability

Outcome

−1   +1

1 bet on red

Probability

Outcome

−10 −8 −6 −4 −2   0   +2 +4 +6 +8 +10

10 bets on red

100 bets on red

Probability

Outcome

−100   −80   −60   −40   −20   0   +20   +40   +60   +80   +100

If the boundary of the universe was simply a point of spacetime, we could keep extending frontiers.

universe was at a jagged edge where space and time were scrunched up and the density was infinite, it would be very difficult to define meaningful boundary conditions.

However, a colleague named Jim Hartle and I realized there was a third possibility. Maybe the universe has no boundary in space and time. At first sight, this seems to be in direct contradiction with the theorems that Penrose and I proved, which showed that the universe must have had a beginning, a boundary in time. However, as explained in Chapter 2, there is another kind of time, called imaginary time, that is at right angles to the ordinary real time that we feel going by. The history of the universe in real time determines its history in imaginary time, and vice versa, but the two kinds of history can be

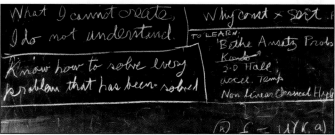 

The blackboard at Caltech at the time of Feynman's death in 1988.   Richard Feynman.

FEYNMAN STORIES

Born in Brooklyn, New York, in 1918, Richard Feynman completed his Ph.D. under John Wheeler at Princeton University in 1942. Shortly afterward, he was drawn into the Manhattan Project. There he was known for both his exuberant personality and practical jokes—at the Los Alamos labs, he enjoyed cracking the top-secret safes—and for being an exceptional physicist: he became a key contributor to atomic bomb theory. Feynman's perpetual curiosity about the world was the very root of his being. It was not only the engine for his scientific success, it led him to numerous astonishing exploits, such as deciphering Mayan hieroglyphics.

In the years following World War II, Feynman found a powerful new way of thinking about quantum mechanics, for which he was awarded the Nobel Prize in 1965. He challenged the basic classical assumption that each particle has one particular history. Instead, he suggested that particles travel from one location to another along every possible path through spacetime. With each trajectory Feynman associated two numbers, one for the size—the amplitude—of a wave and one for its phase—whether it is at a crest or a trough. The probability of a particle going from A to B is found by adding up the waves associated with every possible path that passes through A and B.

Nevertheless, in the everyday world it seems to us that objects follow a single path between their origin and their final destination. This agrees with Feynman's multiple history (or sum-over-histories) idea, because for large objects his rule for assigning numbers to each path ensures that all paths but one cancel out when their contributions are combined. Only one of the infinity of paths matters as far as the motion of macroscopic objects is concerned, and this trajectory is precisely the one emerging from Newton's classical laws of motion.

very different. In particular, the universe need have no beginning or end in imaginary time. Imaginary time behaves just like another direction in space. Thus, the histories of the universe in imaginary time can be thought of as curved surfaces, like a ball, a plane, or a saddle shape, but with four dimensions instead of two (see Fig. 3.9, page 84).

If the histories of the universe went off to infinity like a saddle or a plane, one would have the problem of specifying what the boundary conditions were at infinity. But one can avoid having to specify boundary conditions at all if the histories of the universe in imaginary time are closed surfaces, like the surface of the Earth. The surface of the Earth doesn't have any boundaries or edges. There are no reliable reports of people falling off.

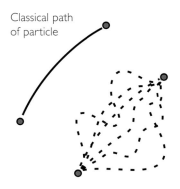

Classical path of particle

In Feynman's path integral a particle takes every possible path.

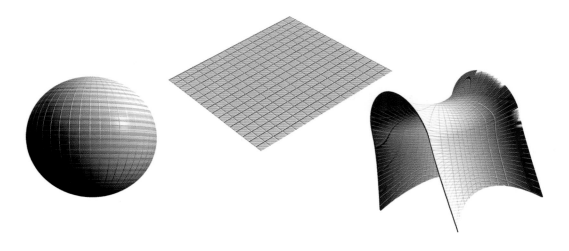

(FIG. 3.9) HISTORIES OF THE UNIVERSE

If the histories of the universe went off to infinity like a saddle, one would have the problem of specifying what the boundary conditions were at infinity. If all the histories of the universe in imaginary time are closed surfaces like that of the Earth, one would not have to specify boundary conditions at all.

EVOLUTION LAWS AND INITIAL CONDITIONS

The laws of physics prescribe how an initial state evolves with time. For instance, if we throw a stone in the air, the laws of gravity will accurately prescribe the stone's subsequent motion.

But we cannot predict where the stone will land exclusively from those laws. For this, we must also know its speed and direction as it left our hand. In other words, we must know the initial conditions —the boundary conditions—of the stone's motion.

Cosmology attempts to describe the evolution of the entire universe using these laws of physics. Hence we must ask what the initial conditions of the universe were to which we should apply these laws.

The initial state may have had a profound impact on basic features of the universe, perhaps even the properties of elementary particles and forces that were crucial for the development of biological life.

One proposal is the *no boundary* condition, the proposal that time and space are finite, forming a closed surface without boundary, just as the surface of the Earth is finite in size but has no boundary. The no boundary proposal is based on Feynman's multiple history idea, but the history of a particle in Feynman's sum is now replaced by a complete spacetime that represents the history of the entire universe. The no boundary condition is precisely the restriction on the possible histories of the universe to those spacetimes that have no boundary in imaginary time. In other words, the boundary condition of the universe is that it has no boundary.

Cosmologists are currently investigating whether initial configurations that are favored by the no boundary proposal, perhaps together with weak anthropic arguments, are likely to evolve to a universe like the one we observe.

If the histories of the universe in imaginary time are indeed closed surfaces, as Hartle and I proposed, it would have fundamental implications for philosophy and our picture of where we came from. The universe would be entirely self-contained; it wouldn't need anything outside to wind up the clockwork and set it going. Instead, everything in the universe would be determined by the laws of science and by rolls of the dice within the universe. This may sound presumptuous, but it is what I and many other scientists believe.

Even if the boundary condition of the universe is that it has no boundary, it won't have just a single history. It will have multiple histories, as suggested by Feynman. There will be a history in imaginary time corresponding to every possible closed surface, and each history in imaginary time will determine a history in real time. Thus we have a superabundance of possibilities for the universe. What picks out the particular universe that we live in from the set of all possible universes? One point we can notice is that many of the possible histories of the universe won't go through the sequence of forming galaxies and stars that was essential to our own development. While it may be that intelligent beings can evolve without galaxies and stars, this seems unlikely. Thus, the very fact that we exist as beings who can ask the question "Why is the universe the way it is?" is a restriction on the history we live in. It implies it is one of the minority of histories that have galaxies and stars. This is an example of what is called the anthropic principle. The anthropic principle says that the universe has to be more or less as we see it, because if it were

The surface of the Earth doesn't have any boundaries or edges. Reports of people falling off are thought to be exaggerations.

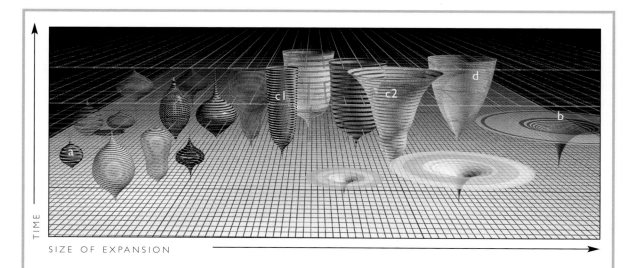

TIME

SIZE OF EXPANSION

THE ANTHROPIC PRINCIPLE

Roughly speaking, the anthropic principle says that we see the universe the way it is, at least in part, because we exist. It is a perspective that is diametrically opposed to the dream of a fully predictive, unified theory in which the laws of nature are complete and the world is the way it is because it could not be otherwise. There are a number of different versions of the anthropic principle, ranging from those that are so weak as to be trivial to those that are so strong as to be absurd. Although most scientists are reluctant to adopt a strong version of the anthropic principle, few people would quarrel with the utility of some weak anthropic arguments.

The weak anthropic principle amounts to an explanation of which of the various possible eras or parts of the universe we *could* inhabit. For instance, the reason why the big bang occurred about ten thousand million years ago is that the universe must be old enough so that some stars will have completed their evolution to produce elements like oxygen and carbon, out of which we are made, and young enough so that some stars would still be providing energy to sustain life.

Within the framework of the no boundary proposal, one can use Feynman's rule for assigning numbers to each history of the universe to find which properties of the universe are likely to occur. In this context, the anthropic principle is implemented by requiring that the histories contain intelligent life. One would feel happier about the anthropic principle, of course, if one could show that a number of different initial configurations for the universe are likely to have evolved to produce a universe like the one we observe. This would imply that the initial state of the part of the universe that we inhabit did not have to be chosen with great care.

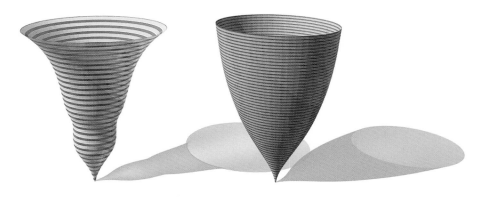

(FIG. 3.10, opposite)
On the far left of the illustration are those universes **(a)** that collapsed on themselves, becoming closed. On the far right are those open universes **(b)** that will continue expanding forever.

Those critical universes that are balanced between falling back on themselves and continuing to expand like **(c1)** or the double inflation of **(c2)** might harbor intelligent life. Our own universe **(d)** is poised to continue expanding for now.

The double inflation could harbor intelligent life.

The inflation of our own universe continues to expand for now.

different, there wouldn't be anyone here to observe it (Fig. 3.10). Many scientists dislike the anthropic principle because it seems rather vague and does not appear to have much predictive power. But the anthropic principle can be given a precise formulation, and it seems to be essential when dealing with the origin of the universe. M-theory, described in Chapter 2, allows a very large number of possible histories for the universe. Most of these histories are not suitable for the development of intelligent life; either they are empty, last for too short a time, are too highly curved, or wrong in some other way. Yet according to Richard Feynman's idea of multiple histories, these uninhabited histories can have quite a high probability (see page 84).

In fact, it doesn't really matter how many histories there may be that don't contain intelligent beings. We are interested only in the subset of histories in which intelligent life develops. This intelligent life need not be anything like humans. Little green aliens would do as well. In fact, they might do rather better. The human race does not have a very good record of intelligent behavior.

As an example of the power of the anthropic principle, consider the number of directions in space. It is a matter of common experience that we live in three-dimensional space. That is to say, we can represent the position of a point in space by three

(FIG. 3.11)

From a distance, a drinking straw looks like a one-dimensional line.

numbers, for example, latitude, longitude, and height above sea level. But why is space three-dimensional? Why isn't it two, or four, or some other number of dimensions, as in science fiction? In M-theory, space has nine or ten dimensions, but it is thought that six or seven of the directions are curled up very small, leaving three dimensions that are large and nearly flat (Fig. 3.11).

Why don't we live in a history in which eight of the dimensions are curled up small, leaving only two dimensions that we notice? A two-dimensional animal would have a hard job digesting food. If it had a gut that went right through it, it would divide the animal in two, and the poor creature would fall apart. So two flat directions are not enough for anything as complicated as intelligent life. On the other hand, if there were four or more nearly flat directions, the gravitational force between two bodies would increase more rapidly as they approached each other. This would mean that planets would not have stable orbits about their suns. They would either fall into the sun (Fig. 3.12A) or escape to the outer darkness and cold (Fig. 3.12B).

FIG. 3.12A

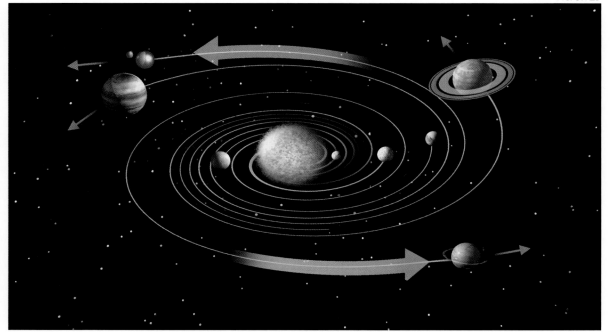

FIG. 3.12B

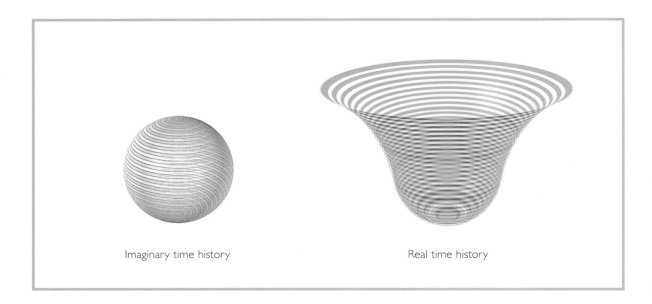

Imaginary time history   Real time history

(FIG. 3.13)

The simplest imaginary time history without boundary is a sphere.

This determines a history in real time that expands in an inflationary manner.

Similarly, the orbits of electrons in atoms would not be stable, so matter as we know it would not exist. Thus, although the idea of multiple histories would allow any number of nearly flat directions, only histories with three flat directions will contain intelligent beings. Only in such histories will the question be asked, "Why does space have three dimensions?"

The simplest history of the universe in imaginary time is a round sphere, like the surface of the Earth, but with two more dimensions (Fig. 3.13). It determines a history of the universe in the real time that we experience, in which the universe is the same at every point of space and is expanding in time. In these respects, it is like the universe we live in. But the rate of expansion is very rapid, and it keeps on getting faster. Such accelerating expansion is called inflation, because it is like the way prices go up and up at an ever-increasing rate.

FIG. 3.14   MATTER ENERGY   GRAVITATION ENERGY

Inflation in prices is generally held to be a bad thing, but in the case of the universe, inflation is very beneficial. The large amount of expansion smoothes out any lumps and bumps there may have been in the early universe. As the universe expands, it borrows energy from the gravitational field to create more matter. The positive matter energy is exactly balanced by the negative gravitational energy, so the total energy is zero. When the universe doubles in size, the matter and gravitational energies both double— so twice zero is still zero. If only the banking world were so simple (Fig. 3.14).

If the history of the universe in imaginary time were a perfectly round sphere, the corresponding history in real time would be a universe that continued to expand in an inflationary manner forever. While the universe is inflating, matter could not fall

(FIG. 3.15) THE INFLATIONARY UNIVERSE

In the hot big bang model, there was not enough time in the early universe for heat to flow from one region to another. Nevertheless we observe that regardless of which direction we look, the temperature of the microwave background radiation is the same. This means that the initial state of the universe must have had exactly the same temperature everywhere.

In an attempt to find a model in which many different initial configurations could have evolved to something like the present universe, it was suggested that the early universe may have been through a period of very rapid expansion. This expansion is said to be inflationary, meaning it takes place at an ever-increasing rate, rather

than the decreasing rate of expansion we observe today. Such an inflationary phase could provide an explanation for the problem of why the universe looks the same in every direction, because there would be enough time for light to travel from one region to another in the early universe.

The corresponding history in imaginary time of a universe that continues to expand in an inflationary manner forever is a perfectly round sphere. But in our own universe, the inflationary expansion slowed down after a fraction of a second, and galaxies could form. In imaginary time, this means that the history of our universe is a sphere with a slightly flattened South Pole.

WHOLESALE PRICE INDEX - INFLATION AND HYPERINFLATION

July 1914	1.0		One German mark in 1914
January 1919	2.6		
July 1919	3.4		Ten thousand marks 1923
January 1920	12.6		
January 1921	14.4		Two million marks 1923
July 1921	14.3		
January 1922	**36.7**		Ten million marks 1923
July 1922	**100.6**		
January 1923	**2,785.0**		One milliard marks 1923
July 1923	**194,000.0**		
November 1923	**726,000,000,000.0**		

together to form galaxies and stars, and life, let alone intelligent life like us, could not develop. Thus although histories of the universe in imaginary time that are perfectly round spheres are allowed by the notion of multiple histories, they are not of much interest. However, histories in imaginary time that are slightly flattened at the south pole of the spheres are much more relevant (Fig. 3.15).

In this case, the corresponding history in real time will expand in an accelerated, inflationary manner at first. But then the expansion will begin to slow down, and galaxies can form. In order for intelligent life to be able to develop, the flattening at the South Pole must be very slight. This will mean that the universe will expand initially by an enormous amount. The record level of monetary inflation occurred in Germany between the world wars, when prices rose billions of times—but the amount of inflation that must have occurred in the universe is at least a billion billion billion times that (Fig. 3.16).

(FIG. 3.16)
INFLATION MAY BE A LAW OF NATURE

Inflation in Germany rose after the peace until by February 1920 the price level was five times as high as it had been in 1918. After July 1922 the phase of hyperinflation began. All confidence in money vanished and the price index rose faster and faster for fifteen months, outpacing the printing presses, which could not produce money as fast as it was depreciating. By late 1923, 300 paper mills were working at top speed and 150 printing companies had 2,000 presses running day and night turning out currency.

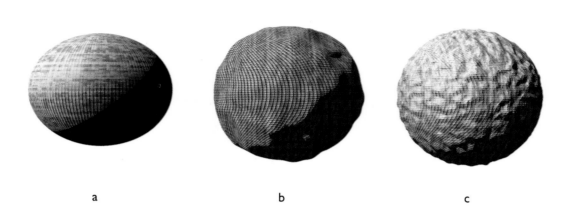

a   b   c

(FIG. 3.17)
PROBABLE AND IMPROBABLE
HISTORIES

Smooth histories like **(a)** are the most probable, but there are only a small number of them.

Although slightly irregular histories **(b)** and **(c)** are each less probable, there are such a large number of them that the likely histories of the universe will have small departures from smoothness.

Because of the uncertainty principle, there won't be just one history of the universe that contains intelligent life. Instead, the histories in imaginary time will be a whole family of slightly deformed spheres, each of which corresponds to a history in real time in which the universe inflates for a long time but not indefinitely. We can then ask which of these allowable histories is the most probable. It turns out that the most probable histories are not completely smooth but have tiny ups and downs (Fig. 3.17). The ripples on the most probable histories really are minuscule. The departures from smoothness are of the order of one part in a hundred thousand. Nevertheless, although they are extremely small, we have managed to observe them as small variations in the microwaves that come to us from different directions in space. The Cosmic Background Explorer satellite was launched in 1989 and made a map of the sky in microwaves.

The different colors indicate different temperatures, but the whole range from red to blue is only about a ten-thousandth of a degree. Yet

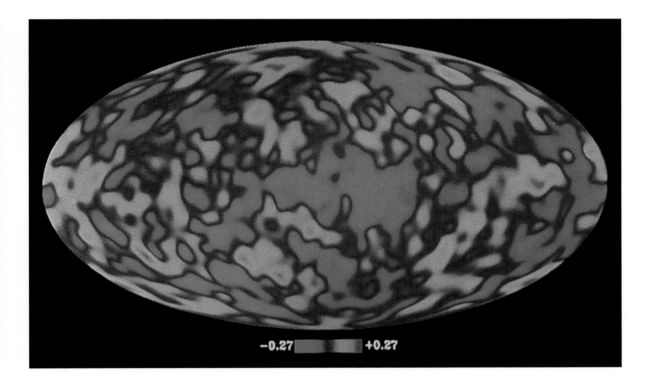

this is enough variation between different regions of the early universe for the extra gravitational attraction in the denser regions to stop them expanding eventually, and to cause them to collapse again under their own gravity to form galaxies and stars. So in principle, at least, the COBE map is the blueprint for all the structures in the universe.

What will be the future behavior of the most probable histories of the universe that are compatible with the appearance of intelligent beings? There seem to be various possibilities, depending on the amount of matter in the universe. If there is more than a certain critical amount, the gravitational attraction between the galaxies will slow them down and will eventually stop them from flying apart. They will then start falling toward each other and will all come together in a big crunch that will be the end of the history of the universe in real time (see Fig. 3.18, page 96).

If the density of the universe is below the critical value, gravity is too weak to stop the galaxies from flying apart forever.

The full sky map made by the COBE satellite DMR instrument, showing evidence for the wrinkles in time.

(FIG. 3.18, above)
One possible end of the universe is the big crunch in which all matter will be sucked back into a vast cataclysmic gravity well.

(FIG. 3.19, opposite)
The long cold whimper in which everything runs down and the last stars flicker out, having exhausted their fuel.

All the stars will burn out, and the universe will get increasingly emptier and colder. So, again, things will come to an end, but in a less dramatic way. Either way, the universe will last a good few billion years more (Fig. 3.19).

As well as matter, the universe may contain what is called "vacuum energy," energy that is present even in apparently empty space. By Einstein's famous equation, $E = mc^2$, this vacuum energy has mass. This means that it has a gravitational effect on the expansion of the universe. But, remarkably enough, the effect of vacuum energy is the opposite of that of matter. Matter causes the expansion to slow down and can eventually stop and reverse it. On the other hand, vacuum energy causes the expansion to accelerate, as in inflation. In fact, vacuum energy acts just like the cosmological constant mentioned in Chapter 1 that Einstein added to his original equations

in 1917, when he realized that they didn't admit a solution representing a static universe. After Hubble's discovery of the expansion of the universe, this motivation for adding a term to the equations disappeared, and Einstein rejected the cosmological constant as a mistake.

However, it may not have been a mistake at all. As described in Chapter 2, we now realize that quantum theory implies that spacetime is filled with quantum fluctuations. In a supersymmetric theory, the infinite positive and negative energies of these ground state fluctuations cancel out between particles of different spin. But we wouldn't expect the positive and negative energies to cancel so completely that there wasn't a small, finite amount of vacuum energy left over, because the universe is not in a supersymmetric state. The only surprise is that the vacuum energy is so nearly zero that it

THE

COSMOLOGICAL

CONSTANT

WAS MY

GREATEST

MISTAKE?

Albert Einstein

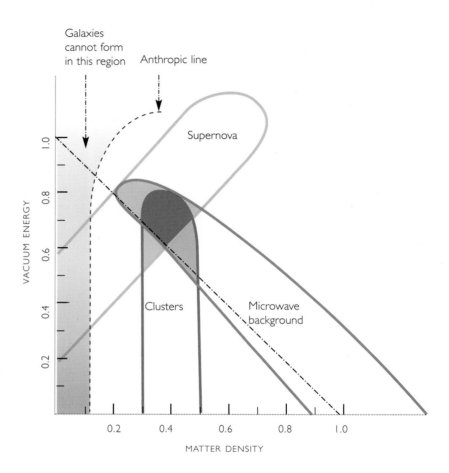

(Fig. 3.20)
By combining observations from distant supernovae, the cosmic microwave background radiation, and the distribution of matter in the universe, the vacuum energy and matter density in the universe can be fairly well estimated.

was not obvious some time ago. Maybe this is another example of the anthropic principle. A history with a larger vacuum energy would not have formed galaxies, so would not contain beings who could ask the question: "Why is the vacuum energy the value we observe?"

We can try to determine the amounts of matter and vacuum energy in the universe from various observations. We can show the results in a diagram in which the matter density is the horizontal direction and vacuum energy is the vertical direction. The dotted line shows the boundary of the region in which intelligent life could develop (Fig. 3.20).

*"I could be bounded in a nutshell
and count myself a king
of infinite space."*

—Shakespeare,
Hamlet, Act 2, Scene 2

Observations of supernovae, clustering, and the microwave background each mark out regions in this diagram. Fortunately, all three regions have a common intersection. If the matter density and vacuum energy lie in this intersection, it means that the expansion of the universe has begun to speed up again, after a long period of slowing down. It seems that inflation may be a law of nature.

In this chapter we have seen how the behavior of the vast universe can be understood in terms of its history in imaginary time, which is a tiny, slightly flattened sphere. It is like Hamlet's nutshell, yet this nut encodes everything that happens in real time. So Hamlet was quite right. We could be bounded in a nutshell and still count ourselves kings of infinite space.

CHAPTER 4

PREDICTING THE FUTURE

How the loss of information in black holes may reduce our ability to predict the future.

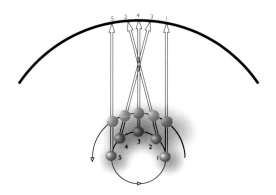

(FIG. 4.1)
An observer on Earth (blue) orbiting the sun watches Mars (red) against a backdrop of constellations.

The complicated apparent motion of the planets in the sky can be explained by Newton's laws and has no influence on personal fortunes.

THE HUMAN RACE HAS ALWAYS WANTED TO CONTROL THE future, or at least to predict what will happen. That is why astrology is so popular. Astrology claims that events on Earth are related to the motions of the planets across the sky. This is a scientifically testable hypothesis, or would be if astrologers stuck their necks out and made definite predictions that could be tested. However, wisely enough, they make their forecasts so vague that they can apply to any outcome. Statements such as "Personal relations may become intense" or "You will have a financially rewarding opportunity" can never be proved wrong.

But the real reason most scientists don't believe in astrology is not scientific evidence or the lack of it but because it is not consistent with other theories that have been tested by experiment. When Copernicus and Galileo discovered that the planets orbit the Sun rather than the Earth, and Newton discovered the laws that govern their motion, astrology became extremely implausible. Why should the positions of other planets against the background sky as seen from Earth have any correlations with the macromolecules on a minor planet that call themselves intelligent life (Fig. 4.1)? Yet this is what astrology would have us believe. There is no more experimental evidence for some of the theories described in this book

"Mars occupies Sagittarius this month and for you it will be a time to seek self-knowledge. Mars asks you to live life according to what feels right as opposed to what others think is right. And this will happen.

On the 20th Saturn arrives in the area of your solar chart related to commitment and career and you will be learning to take responsibilities and deal with difficult relationships.

However, at the time of the full moon you will gain a wonderful insight and overview of your entire life which will transform you."

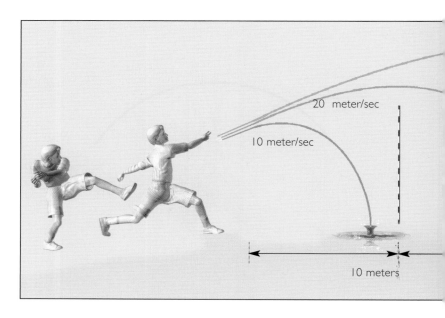

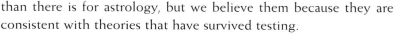

20 meter/sec

10 meter/sec

10 meters

(FIG. 4.2)

If you know where and at what speed a baseball is thrown, you can predict where it will go.

(FIG. 4.3)

than there is for astrology, but we believe them because they are consistent with theories that have survived testing.

The success of Newton's laws and other physical theories led to the idea of scientific determinism, which was first expressed at the beginning of the nineteenth century by the French scientist the Marquis de Laplace. Laplace suggested that if we knew the positions and velocities of all the particles in the universe at one time, the laws of physics should allow us to predict what the state of the universe would be at any other time in the past or in the future (Fig. 4.2).

In other words, if scientific determinism holds, we should in principle be able to predict the future and wouldn't need astrology. Of course, in practice even something as simple as Newton's theory of gravity produces equations that we can't solve exactly for more than two particles. Furthermore, the equations often have a property known as chaos, so that a small change in position or velocity at one time can lead to completely different behavior at later times. As those who have seen *Jurassic Park* know, a tiny disturbance in one place can cause a major change in another. A butterfly flapping its wings in Tokyo can cause rain in New York's Central Park (Fig. 4.3). The trouble is the sequence of events is not repeatable. The next time the butterfly flaps its wings, a host of other factors will be different

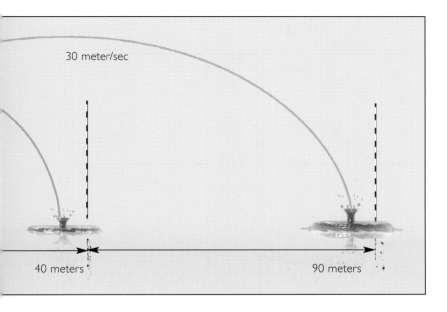

30 meter/sec

40 meters

90 meters

and will also influence the weather. That is why weather forecasts are so unreliable.

Thus, although in principle the laws of quantum electrodynamics should allow us to calculate everything in chemistry and biology, we have not had much success in predicting human behavior from mathematical equations. Nevertheless, despite these practical difficulties most scientists have comforted themselves with the idea that, again in principle, the future is predictable.

IN

?!

OUT

At first sight, determinism would also seem to be threatened by the uncertainty principle, which says that we cannot measure accurately both the position and the velocity of a particle at the same time. The more accurately we measure the position, the less accurately we can determine the velocity, and vice versa. The Laplace version of scientific determinism held that if we knew the positions and velocities of particles at one time, we could determine their positions and velocities at any time in the past or future. But how could we even get started if the uncertainty principle prevented us from knowing accurately both the positions and the velocities at one time? However good our computer is, if we put lousy data in, we will get lousy predictions out.

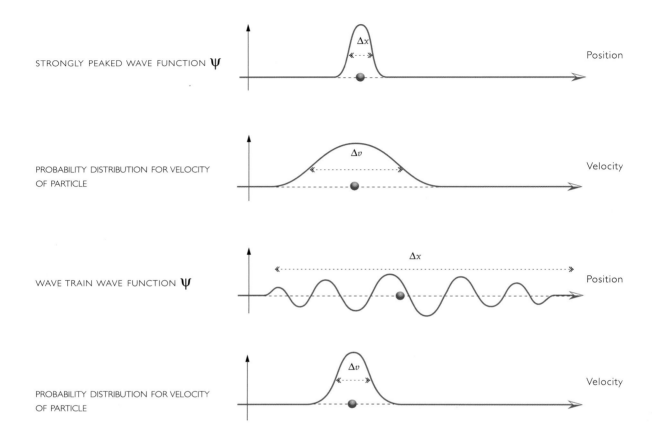

STRONGLY PEAKED WAVE FUNCTION ψ

PROBABILITY DISTRIBUTION FOR VELOCITY OF PARTICLE

WAVE TRAIN WAVE FUNCTION ψ

PROBABILITY DISTRIBUTION FOR VELOCITY OF PARTICLE

(FIG. 4.4)

The wave function determines the probabilities that the particle will have different positions and velocities in such a way that Δx and Δv obey the uncertainty principle.

However, determinism *was* restored in a modified form in a new theory called quantum mechanics, which incorporated the uncertainty principle. In quantum mechanics, one can, roughly speaking, accurately predict half of what one would expect to predict in the classical Laplace point of view. In quantum mechanics, a particle does not have a well-defined position or velocity, but its state *can* be represented by what is called a wave function (Fig. 4.4).

A wave function is a number at each point of space that gives the probability that the particle is to be found at that position. The rate at which the wave function changes from point to point tells how probable different particle velocities are. Some wave functions are sharply peaked at a particular point in space. In these cases, there is only a small amount of uncertainty in the position of the particle. But we can also see in the diagram that in such cases, the wave function changes rapidly near the point, up on one side and

down on the other. That means the probability distribution for the velocity is spread over a wide range. In other words, the uncertainty in the velocity is large. Consider, on the other hand, a continuous train of waves. Now there is a large uncertainty in position but a small uncertainty in velocity. So the description of a particle by a wave function does not have a well-defined position or velocity. It satisfies the uncertainty principle. We now realize that the wave function is *all* that can be well defined. We cannot even suppose that the particle has a position and velocity that are known to God but are hidden from us. Such "hidden-variable" theories predict results that are not in agreement with observation. Even God is bound by the uncertainty principle and cannot know the position and velocity; He can only know the wave function.

The rate at which the wave function changes with time is given by what is called the Schrödinger equation (Fig. 4.5). If we know the

(FIG. 4.5)
THE SCHRÖDINGER EQUATION

The evolution in time of the wave function ψ is determined by the Hamiltonian operator H, which is associated with the energy of the physical system under consideration.

(Fig. 4.6)
In the flat spacetime of special relativity observers moving at different speeds will have different measures of time, but we can use the Schrödinger equation in any of these times to predict what the wave function will be in the future.

wave function at one time, we can use the Schrödinger equation to calculate it at any other time, past or future. Therefore, there is still determinism in quantum theory, but it is on a reduced scale. Instead of being able to predict both the positions and the velocities, we can predict only the wave function. This can allow us to predict either the positions or the velocities, but not both accurately. Thus in quantum theory the ability to make exact predictions is just half what it was in the classical Laplace worldview. Nevertheless, within this restricted sense it is still possible to claim that there is determinism.

However, the use of the Schrödinger equation to evolve the wave function forward in time (that is, to predict what it will be at future times) implicitly assumes that time runs on smoothly everywhere, forever. This was certainly true in Newtonian physics. Time was assumed to be absolute, meaning that each event in the history of the universe was labeled by a number called time, and that a series of time labels ran smoothly from the infinite past to the infinite future. This is what might be called the commonsense view of time, and it is the view of time that most people and even most physicists have at the back of their minds. However, in 1905, as we have seen, the concept of absolute time was overthrown by the special theory of relativity, in which time was no longer an independent quantity on its own but was just one direction in a four-dimensional continuum called spacetime. In special relativity, different observers traveling at

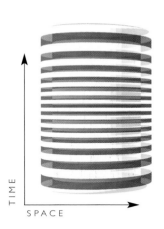

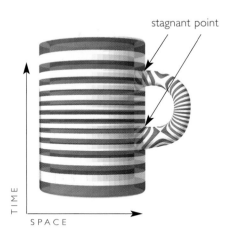

stagnant point

TIME

SPACE

TIME

SPACE

different velocities move through spacetime on different paths. Each observer has his or her own measure of time along the path he or she is following, and different observers will measure different intervals of time between events (Fig. 4.6).

Thus in special relativity there is no unique absolute time that we can use to label events. However, the spacetime of special relativity is flat. This means that in special relativity, the time measured by any freely moving observer increases smoothly in spacetime from minus infinity in the infinite past to plus infinity in the infinite future. We can use any of these measures of time in the Schrödinger equation to evolve the wave function. In special relativity, therefore, we still have the quantum version of determinism.

The situation was different in the general theory of relativity, in which spacetime was not flat but curved, and distorted by the matter and energy in it. In our solar system, the curvature of spacetime is so slight, at least on a macroscopic scale, that it doesn't interfere with our usual idea of time. In this situation, we could still use this time in the Schrödinger equation to get a deterministic evolution of the wave function. However, once we allow spacetime to be curved, the door is opened to the possibility that it may have a structure that doesn't admit a time that increases smoothly for every observer, as we would expect for a reasonable measure of time. For example, suppose that spacetime was like a vertical cylinder (Fig. 4.7).

(FIG. 4.7) TIME STANDS STILL

A measure of time would necessarily have stagnation points where the handle joined the main cylinder: points where time stood still. At these points, time would not increase in any direction. Therefore, one could not use the Schrödinger equation to predict what the wave function will be in the future.

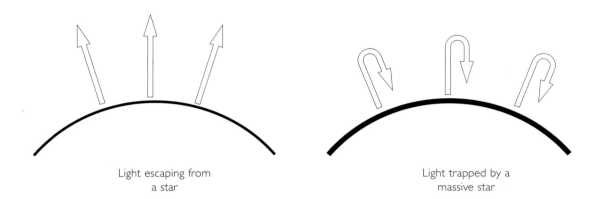

Light escaping from
a star

Light trapped by a
massive star

FIG. 4.9

FIG. 4.8

Height up the cylinder would be a measure of time that increased for every observer and ran from minus infinity to plus infinity. However, imagine instead that spacetime was like a cylinder with a handle (or "wormhole") that branched off and then joined back on. Then any measure of time would necessarily have stagnation points where the handle joined the main cylinder: points where time stood still. At these points, time would not increase for any observer. In such a spacetime, we could not use the Schrödinger equation to get a deterministic evolution for the wave function. Watch out for wormholes: you never know what may come out of them.

Black holes are the reason we think time will not increase for every observer. The first discussion of black holes appeared in 1783. A former Cambridge don, John Michell, presented the following argument. If one fires a particle, such as a cannonball, vertically upward, its ascent will be slowed by gravity, and eventually the particle will stop moving upward and will fall back (Fig. 4.8). However, if the initial upward velocity is greater than a critical value called the escape velocity, gravity will never be strong enough to stop the particle, and it will get away. The escape velocity is about 12 kilometers per second for the Earth, and about 618 kilometers per second for the Sun.

THE SCHWARZSCHILD BLACK HOLE

In 1916 the German astronomer Karl Schwarzschild found a solution to Einstein's theory of relativity that represents a spherical black hole. Schwarzschild's work revealed a stunning implication of general relativity. He showed that if the mass of a star is concentrated in a small enough region, the gravitational field at the surface of the star becomes so strong that even light can no longer escape. This is what we now call a black hole, a region of spacetime bounded by a so-called event horizon, from which it is impossible for anything, including light, to reach a distant observer.

For a long time most physicists, including Einstein, were skeptical whether such extreme configurations of matter could actually ever occur in the real universe. However, we now understand that when any sufficiently heavy nonrotating star, however complicated its shape and internal structure, runs out of nuclear fuel, it will necessarily collapse to a perfectly spherical Schwarzschild black hole. The radius (R) of the black hole's event horizon depends only on its mass; it is given by the formula:

$$R = \frac{2GM}{c^2}$$

In this formula the symbol (c) stands for the speed of light, (G) for Newton's constant, and (M) for the mass of the black hole. A black hole with the same mass as the Sun, for instance, would have a radius of only two miles!

Both of these escape velocities are much higher than the speed of real cannonballs, but they are small compared to the speed of light, which is 300,000 kilometers per second. Thus, light can get away from the Earth or Sun without much difficulty. However, Michell argued that there could be stars that are much more massive than the Sun and have escape velocities greater than the speed of light (Fig. 4.9). We would not be able to see these stars, because any light they sent out would be dragged back by the gravity of the star. Thus they would be what Michell called dark stars and we now call black holes.

Michell's idea of dark stars was based on Newtonian physics, in which time was absolute and went on regardless of what happened. Thus they didn't affect our ability to predict the future in the classical Newtonian picture. But the situation was very different in the general theory of relativity, in which massive bodies curve spacetime.

In 1916, shortly after the theory was first formulated, Karl Schwarzschild (who died soon after of an illness contracted on the Russian front in the First World War) found a solution of the field equations of general relativity that represented a black hole. What Schwarzschild had found wasn't understood or its importance recognized for many years. Einstein himself never believed in black

(Fig. 4.10)

The quasar 3C273, the first quasi-stellar radio source to be discovered, produces a large amount of power in a small region. Matter falling into a black hole seems to be the only mechanism that can account for such a high luminosity.

JOHN WHEELER

John Archibald Wheeler was born in 1911 in Jacksonville, Florida. He earned his Ph.D. from Johns Hopkins University in 1933 for his work on the scattering of light by the helium atom. In 1938 he worked with the Danish physicist Niels Bohr to develop the theory of nuclear fission. For a while thereafter Wheeler, together with his graduate student Richard Feynman, concentrated on the study of electrodynamics; but shortly after America entered World War II, both went on to contribute to the Manhattan Project.

In the early 1950s, inspired by Robert Oppenheimer's work in 1939 on the gravitational collapse of a massive star, Wheeler turned his attention to Einstein's theory of general relativity. At that time most physicists were caught up in the study of nuclear physics and general relativity was not really regarded as relevant to the physical world. But almost single-handedly Wheeler transformed the field, both through his research and through his teaching of Princeton's first course on relativity.

Much later, in 1969, he coined the term *black hole* for the collapsed state of matter, which few yet believed was real. Inspired by the work of Werner Israel, he conjectured that black holes have no hair, which meant that the collapsed state of any nonrotating massive star could in fact be described by Schwarzschild's solution.

holes, and his attitude was shared by most of the old guard in general relativity. I remember going to Paris to give a seminar on my discovery that quantum theory means that black holes aren't completely black. My seminar fell rather flat because at that time almost no one in Paris believed in black holes. The French also felt that the name as they translated it, *trou noir,* had dubious sexual connotations and should be replaced by *astre occlu,* or "hidden star." However, neither this nor other suggested names caught the public imagination like the term *black hole,* which was first introduced by John Archibald Wheeler, the American physicist who inspired much of the modern work in this field.

The discovery of quasars in 1963 brought forth an outburst of theoretical work on black holes and observational attempts to detect them (Fig. 4.10). Here is the picture that has emerged. Consider what we believe would be the history of a star with a mass twenty times that of the Sun. Such stars form from clouds of gas, like those in the Orion Nebula (Fig. 4.11). As clouds of gas contract under their own gravity, the gas heats up and eventually becomes hot enough to start the nuclear fusion reaction that converts hydrogen into helium. The heat generated by this process creates a pressure that supports the star against its own gravity and stops it from contracting further. A star will stay in this state for a long time, burning hydrogen and radiating light into space.

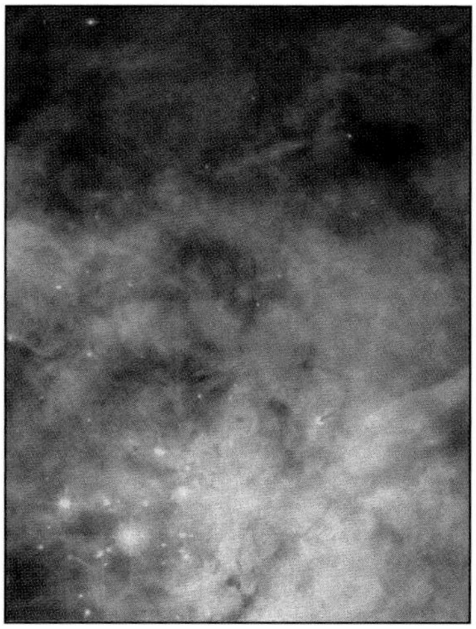

(FIG. 4.11)
Stars form in clouds of gas and dust like the Orion Nebula.

The gravitational field of the star will affect the paths of light rays coming from it. One can draw a diagram with time plotted upward and distance from the center of the star plotted horizontally (see Fig. 4.12, page 114). In this diagram, the surface of the star is represented by two vertical lines, one on either side of the center. One can choose that time be measured in seconds and distance in light-seconds—the distance light travels in a second. When we use these units, the speed of light is 1; that is, the speed of light is 1 light-second per second. This means that far from the star and its gravitational field, the path of a light ray on the diagram is a line at a 45-degree angle to the vertical. However, nearer the star, the curvature of space-time produced by the mass of the star will change the paths of the light rays and cause them to be at a smaller angle to the vertical.

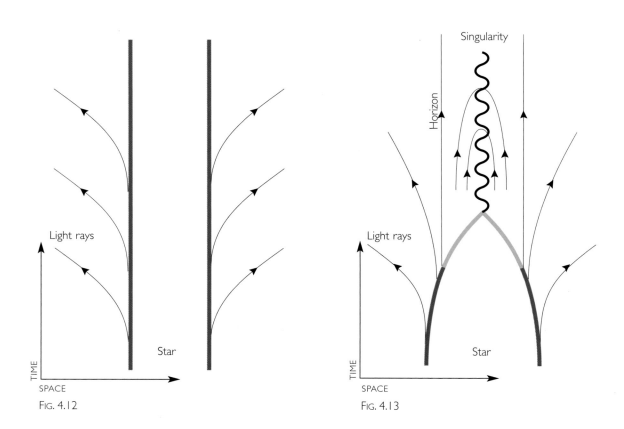

FIG. 4.12

FIG. 4.13

(FIG. 4.12) Spacetime around a non-collapsing star. Light rays can escape from the surface of the star (the red vertical lines). Far from the star, the light rays are at 45 degrees to the vertical, but near the star the warping of spacetime by the mass of the star causes light rays to be at a smaller angle to the vertical.

(FIG. 4.13) If the star collapses (the red lines meeting at a point) the warping becomes so large that light rays near the surface move inward. A black hole is formed, a region of spacetime from which it is not possible for light to escape.

Massive stars will burn their hydrogen into helium much faster than the Sun does. This means they can run out of hydrogen in as little as a few hundred million years. After that, such stars face a crisis. They can burn their helium into heavier elements such as carbon and oxygen, but these nuclear reactions do not release much energy, so the stars lose heat and the thermal pressure that supports them against gravity. Therefore they begin to get smaller. If they are more than about twice the mass of the Sun, the pressure will never be sufficient to stop the contraction. They will collapse to zero size and infinite density to form what is called a singularity (Fig. 4.13). In the diagram of time against distance from the center, as a star shrinks, the paths of light rays from its surface will start out

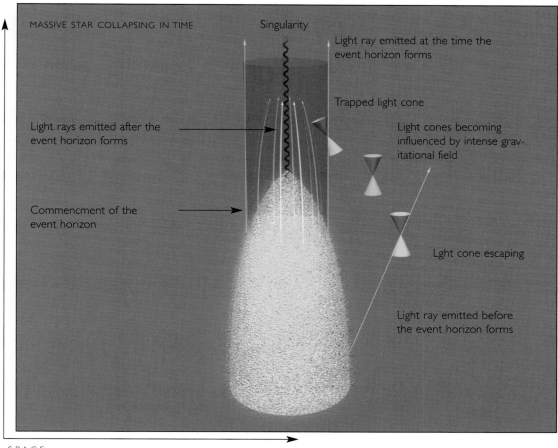

MASSIVE STAR COLLAPSING IN TIME

Singularity

Light ray emitted at the time the event horizon forms

Trapped light cone

Light rays emitted after the event horizon forms

Light cones becoming influenced by intense gravitational field

Commencment of the event horizon

Lght cone escaping

Light ray emitted before the event horizon forms

TIME

SPACE

at smaller and smaller angles to the vertical. When the star reaches a certain critical radius, the path will be vertical on the diagram, which means that the light will hover at a constant distance from the center of the star, never getting away. This critical path of light will sweep out a surface called the event horizon, which separates the region of spacetime from which light can escape from the region from which it cannot. Any light emitted by the star after it passes the event horizon will be bent back inward by the curvature of spacetime. The star will have become one of Michell's dark stars, or, as we say now, a black hole.

How can you detect a black hole if no light can get out of it? The answer is that a black hole still exerts the same gravitational

The horizon, the outer boundary of a black hole, is formed by light rays that just fail to get away from the black hole, but stay hovering at a constant distance from the center.

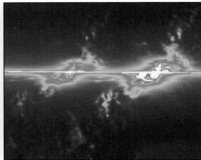

(FIG. 4.15)

A BLACK HOLE AT THE CENTER OF A GALAXY

Left: The galaxy NGC 4151 revealed by the wide-field planetary camera.

Center: The horizontal line passing through the image is from light generated by the black hole at the center of 4151.

Right: Image showing the velocity of oxygen emissions. All the evidence indicates that NGC 4151 contains a black hole about a hundred million times the mass of the Sun.

(FIG. 4.14)

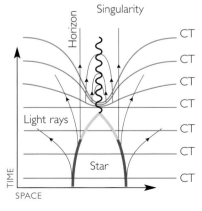

(CT—Lines of constant time)

pull on neighboring objects as did the body that collapsed. If the Sun were a black hole and had managed to become one without losing any of its mass, the planets would still orbit as they do now.

One way of searching for a black hole is therefore to look for matter that is orbiting what seems to be an unseen compact massive object. A number of such systems have been observed. Perhaps the most impressive are the giant black holes that occur in the centers of galaxies and quasars (Fig. 4.15).

The properties of black holes that have been discussed thus far don't raise any great problems with determinism. Time will come to an end for an astronaut who falls into a black hole and hits the singularity. However, in general relativity, one is free to measure time at different rates in different places. One could therefore speed up the astronaut's watch as he or she approached the singularity, so that it still registered an infinite interval of time. On the time-and-distance diagram (Fig. 4.14), the surfaces of constant values of this new time would be all crowded together at the center, below the point where the singularity appeared. But they would agree with the usual measure of time in the nearly flat spacetime far away from the black hole.

One could use this time in the Schrödinger equation and calculate the wave function at later times if one knew it initially. Thus one still has determinism. It is worth noting, however, that at late times, part of the wave function is inside the black hole, where it can't be observed by someone outside. Thus an observer who is sensible enough not to fall into a black hole cannot run the Schrödinger equation backward and calculate the wave function at early times. To do that, he or she would need to know the part of the wave function that is inside the black hole. This contains the

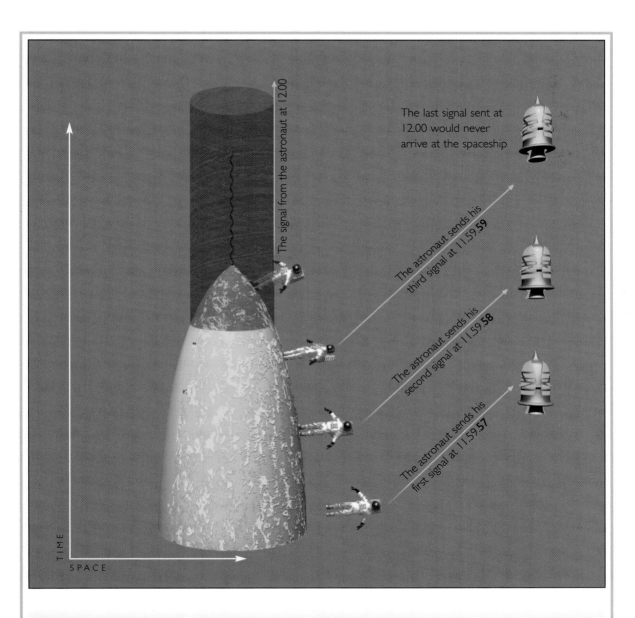

The signal from the astronaut at 12.00

The last signal sent at 12.00 would never arrive at the spaceship

The astronaut sends his third signal at 11.59.59

The astronaut sends his second signal at 11.59.58

The astronaut sends his first signal at 11.59.57

TIME

SPACE

The illustration above shows an astronaut who lands on a collapsing star at 11.59.57 and joins the star as it shrinks below the critical radius where gravity is so strong that no signal can escape. He sends signals from his watch to a spaceship orbiting the star at regular intervals.

Someone watching the star at a distance will never see it cross the event horizon and enter the black hole. Instead, the star will appear to hover just outside the critical radius, and a clock on the surface of the star will seem to slow down and stop.

The no-hair result.

BLACK HOLE TEMPERATURE

A black hole emits radiation as if it were a hot body with a temperature (T) that depends only on its mass. More precisely, the temperature is given by the following formula:

$$T = \frac{\hbar c^3}{8\pi\, k\, G\, M}$$

In this formula the symbol (c) stands for the speed of light, ($\hbar$) for Planck's constant, (G) for Newton's gravitational constant, and (k) for Boltzmann's constant.

Finally, (M) represents the mass of the black hole, so the smaller the black hole, the higher its temperature. This formula tells us that the temperature of a black hole of a few solar masses is only about a millionth of a degree above absolute zero.

information about what fell into the hole. This is potentially a very large amount of information, because a black hole of a given mass and rate of rotation can be formed from a very large number of different collections of particles; a black hole does not depend on the nature of the body that had collapsed to form it. John Wheeler called this result "a black hole has no hair." For the French, this just confirmed their suspicions.

The difficulty with determinism arose when I discovered that black holes aren't completely black. As we saw in Chapter 2, quantum theory means that fields can't be exactly zero even in what is called the vacuum. If they were zero, they would have both an exact value or position at zero and an exact rate of change or velocity that was also zero. This would be a violation of the uncertainty principle, which says that the position and velocity can't both be well defined. All fields must instead have a certain amount of what are called vacuum fluctuations (in the same way that the pendulum in Chapter 2 had to have zero point fluctuations). Vacuum fluctuations can be interpreted in several ways that seem different but are in fact mathematically equivalent. From a positivist viewpoint, one is free to use whatever picture is most useful for the problem in question. In this case it is helpful to think of vacuum fluctuations as pairs of virtual particles that appear together at some point of spacetime, move apart, and come back together and annihilate each other. "Virtual" means that these particles cannot be observed directly, but their indirect effects *can* be measured, and they agree with theoretical predictions to a remarkable degree of accuracy (Fig. 4.16).

If a black hole is present, one member of a pair of particles may fall into the black hole, leaving the other member free to escape to infinity (Fig. 4.17). To someone far from the black hole, the escaping particles appear to have been radiated by the black hole. The spectrum of a black hole is exactly what we would expect from a hot body, with a temperature proportional to the gravitational field on the horizon—the boundary—of the black hole. In other words, the temperature of a black hole depends on its size.

A black hole of a few solar masses would have a temperature of about a millionth of a degree above absolute zero, and a larger black hole would have an even lower temperature. Thus any quantum

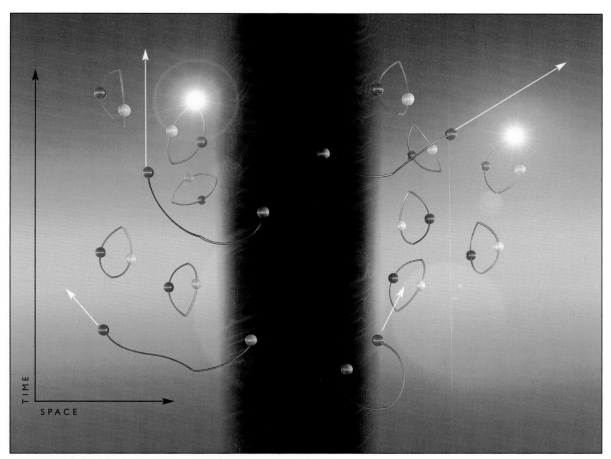

TIME

SPACE

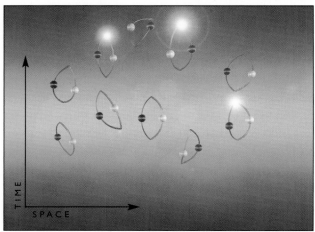

TIME

SPACE

(Fig. 4.17)
Above: Virtual particles appearing and annihilating one another, close to the event horizon of a black hole.

One of the pair falls into the black hole while its twin is free to escape. From outside the event horizon it appears that the black hole is radiating the particles that escape.

(Fig. 4.16)
Left: In empty space particle pairs appear, lead a brief existence, and then annihilate one another.

Events that will never be seen by the observer

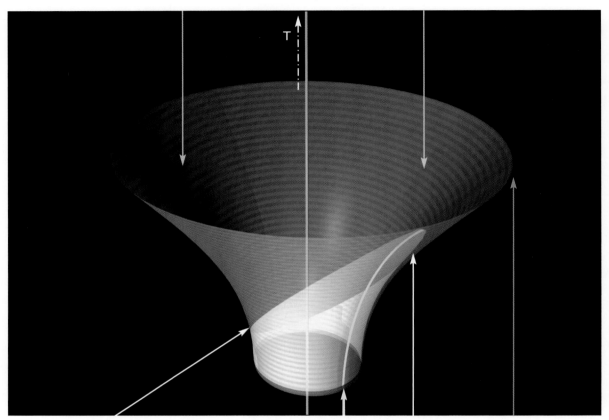

Observer's event horizon   History of observer   Observer's event horizon   Surface of constant time

(FIG. 4.18)

The de Sitter solution of the field equations of general relativity represents a universe that expands in an inflationary manner. In the diagram time is shown in the upward and the size of the universe in the horizontal direction. Spatial distances increase so rapidly that light from distant galaxies never reaches us, and there is an event horizon, a boundary of the region we cannot observe, as in a black hole.

radiation from such black holes would be utterly swamped by the 2.7-degree radiation left over from the hot big bang—the cosmic background radiation that we discussed in Chapter 2. It would be possible to detect the radiation from much smaller and hotter black holes, but there don't seem to be many of them around. That is a pity. If one were discovered, I would get a Nobel Prize. However, we have indirect observational evidence for this radiation, and that evidence comes from the early universe. As described in Chapter 3, it is thought that very early in its history, the universe went through an inflationary period during which it expanded at an ever-increasing rate. The expansion during this period would have been so rapid that some objects would be too distant from us for their light ever to reach us; the universe would have expanded too much and too rapidly while that light was traveling toward us. Thus there would be a horizon in the universe like the horizon of a black hole, separating the region from which light can reach us and the region from which it cannot (Fig. 4.18).

Very similar arguments show that there should be thermal radiation from this horizon, as there is from a black hole horizon. In thermal radiation, we have learned to expect a characteristic spectrum of density fluctuations. In this case, these density fluctuations would have expanded with the universe. When their length scale became longer than the size of the event horizon, they would have become frozen in, so that we can observe them today as small variations in the temperature of the cosmic background radiation left over from the early universe. The observations of those variations agree with the predictions of thermal fluctuations with remarkable accuracy.

Even if the observational evidence for black hole radiation is a bit indirect, everyone who has studied the problem agrees it must occur in order to be consistent with our other observationally tested theories. This has important implications for determinism. The radiation from a black hole will carry away energy, which must mean that the black hole will lose mass and get smaller. In turn, this will mean that its temperature will rise and the rate of radiation will increase. Eventually the black hole will get down to zero mass. We don't know how to calculate what happens at this point, but the only natural, reasonable outcome would seem to be that the black hole disappears completely. So what happens then to the part of the wave function inside the black hole and the information it contains about what had fallen into the black hole? The first guess might be that this part of the wave function, and the information it carries, would emerge when the black hole finally disappears. However, information cannot be carried for free, as one realizes when one gets a telephone bill.

Information requires energy to carry it, and there's very little energy left in the final stages of a black hole. The only plausible way the information

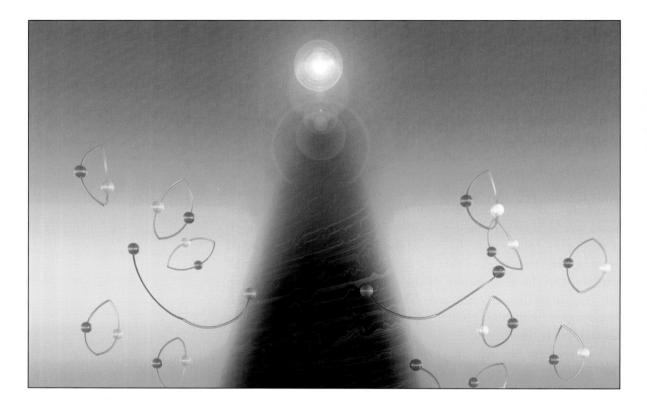

(FIG. 4.19)

The positive energy carried away by the thermal radiation from its horizon reduces the black hole's mass. As it loses mass, the temperature of the black hole rises and its rate of radiation increases, so it loses mass more and more quickly. We don't know what happens if the mass becomes extremely small, but the most likely outcome seems to be that the black hole would disappear completely.

inside could get out would be if it emerged continuously with the radiation, rather than waiting for this final stage. However, according to the picture of one member of a virtual-particle pair falling in and the other member escaping, one would not expect the escaping particle to be related to what fell in, or to carry away information about it. So the only answer would seem to be that the information in the part of the wave function inside the black hole gets lost (Fig. 4.19).

Such loss of information would have important implications for determinism. To start with, we have noted that even if you knew the wave function after the black hole disappeared, you could not run the Schrödinger equation backward and calculate what the wave function was before the black hole formed. What that was would depend in part on the bit of the wave function that got lost in the black hole. We are used to thinking we can know the past exactly. However, if information gets lost in black holes, this is not the case. Anything could have happened.

In general, however, people such as astrologers and those who consult them are more interested in predicting the future than in retrodicting the past. At first glance, it might seem that the loss of part of the wave function down the black hole would not prevent us from predicting the wave function outside the black hole. But it turns out that this loss does interfere with such a prediction, as we can see when we consider a thought experiment proposed by Einstein, Boris Podolsky, and Nathan Rosen in the 1930s.

Imagine that a radioactive atom decays and sends out two particles in opposite directions and with opposite spins. An observer who looks only at one particle cannot predict whether it will be spinning to the right or left. But if the observer measures it to be spinning to the right, then he or she can predict with certainty that the other particle will be spinning to the left, and vice versa (Fig. 4.20). Einstein thought that this proved that quantum theory was ridiculous: the other particle might be at the other side of the

(FIG. 4.20)

In the Einstein-Podolsky-Rosen thought experiment, the observer who has measured the spin of one particle will know the direction of the spin of the second particle.

123

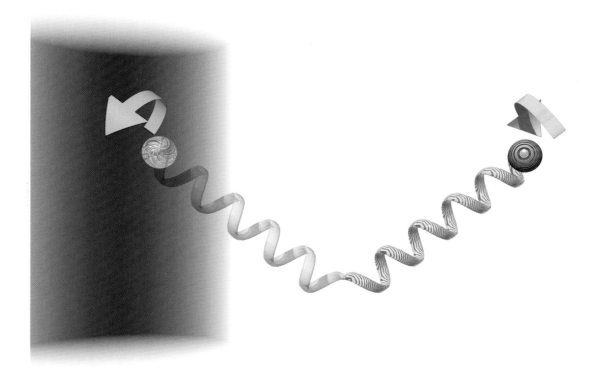

(FIG. 4.21)

A virtual-particle pair has a wave function that predicts that both particles will have opposite spins. But if one particle falls into the black hole, it is impossible to predict with certainty the spin of the remaining particle.

galaxy by now, yet one would instantaneously know which way it was spinning. However, most other scientists agree that it was Einstein who was confused, not quantum theory. The Einstein-Podolsky-Rosen thought experiment does not show that one is able to send information faster than light. That would be the ridiculous part. One cannot *choose* that one's own particle will be measured to be spinning to the right, so one cannot prescribe that the distant observer's particle should be spinning to the left.

In fact, this thought experiment is exactly what happens with black hole radiation. The virtual-particle pair will have a wave function that predicts that the two members will definitely have opposite spins (Fig. 4.21). What we would like to do is predict the spin and wave function of the outgoing particle, which we could do if we could observe the particle that has fallen in. But that particle is now inside the black hole, where its spin and wave function cannot be

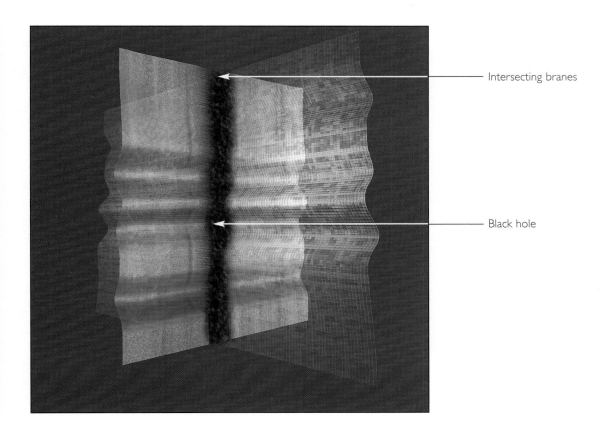

Intersecting branes

Black hole

measured. Because of this, it is not possible to predict the spin or the wave function of the particle that escapes. It can have different spins and different wave functions, with various probabilities, but it doesn't have a unique spin or wave function. Thus it would seem that our power to predict the future would be further reduced. The classical idea of Laplace, that one could predict both the positions and the velocities of particles, had to be modified when the uncertainty principle showed that one could not accurately measure both positions and velocities. However, one could still measure the wave function and use the Schrödinger equation to predict what it should be in the future. This would allow one to predict with certainty one combination of position and velocity—which is half of what one could predict according to Laplace's ideas. We can predict with certainty that the particles have opposite spins, but if one particle falls into the black hole, there is no prediction we can make with certainty about the

(FIG. 4.22)
Black holes can be thought of as the intersections of p-branes in the extra dimensions of spacetime. Information about the internal states of black holes would be stored as waves on the p-branes.

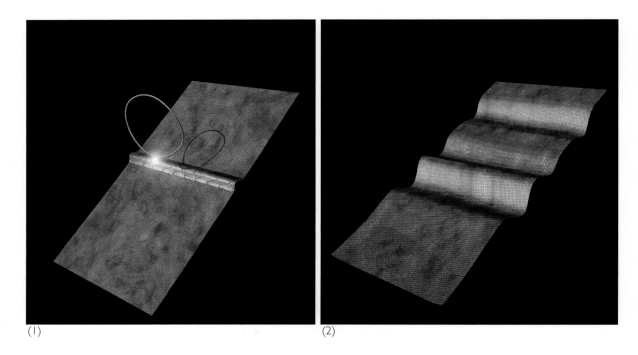

(1)

(2)

(FIG. 4.23)
A particle falling into a black hole can be thought of as a closed loop of string hitting a p-brane **(1)**. It will excite waves in the p-brane **(2)**. Waves can come together and cause part of the p-brane to break off as a closed string **(3)**. This would be a particle emitted by the black hole.

remaining particle. This means that there isn't *any* measurement outside the black hole that can be predicted with certainty: our ability to make definite predictions would be reduced to zero. So maybe astrology is no worse at predicting the future than the laws of science.

Many physicists didn't like this reduction in determinism and therefore suggested that information about what is inside can somehow get out of a black hole. For years it was just a pious hope that some way to save the information would be found. But in 1996 Andrew Strominger and Cumrun Vafa made an important advance. They chose to regard a black hole as being made up of a number of building blocks, called p-branes (see page 54).

Recall that one way of thinking about p-branes is as sheets that move through the three dimensions of space and also through seven extra dimensions that we don't notice (see Fig. 4.22, page 125). In certain cases, one can show that the number of waves on the p-branes is the same as the amount of information one would expect the black hole to contain. If particles hit the p-branes, they excite extra waves on the branes. Similarly, if waves moving in different

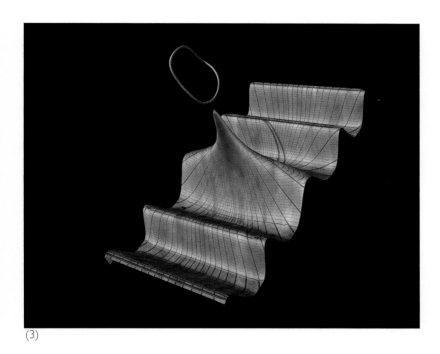

(3)

directions on the p-branes come together at some point, they can create a peak so great that a bit of the p-brane breaks away and goes off as a particle. Thus the p-branes can absorb and emit particles like black holes (Fig. 4.23).

One can regard the p-branes as an effective theory; that is, while we don't need to believe that there actually are little sheets moving through a flat spacetime, black holes can behave as if they were made up of such sheets. It is like water, which is made up of billions and billions of H_2O molecules with complicated interactions. But a smooth fluid is a very good effective model. The mathematical model of black holes as made of p-branes gives results similar to the virtual-particle pair picture described earlier. Thus from a positivist viewpoint, it is an equally good model, at least for certain classes of black hole. For these classes, the p-brane model predicts exactly the same rate of emission that the virtual-particle pair model predicts. However, there is one important difference: in the p-brane model, information about what falls into the black hole will be stored in the wave function for the waves on the p-branes. The p-branes are

regarded as sheets in *flat* spacetime, and for that reason, time will flow forward smoothly, the paths of light rays won't be bent, and the information in the waves won't be lost. Instead, the information will eventually emerge from the black hole in the radiation from the p-branes. Thus, according to the p-brane model, we can use the Schrödinger equation to calculate what the wave function will be at later times. Nothing will get lost, and time will roll smoothly on. We will have complete determinism in the quantum sense.

So which of these pictures is correct? Does part of the wave function get lost down black holes, or does all the information get out again, as the p-brane model suggests? This is one of the outstanding questions in theoretical physics today. Many people believe that recent work shows that information is not lost. The world is safe and predictable, and nothing unexpected will happen. But it's not clear. If one takes Einstein's general theory of relativity seriously, one must allow the possibility that spacetime ties itself in a knot and information gets lost in the folds. When the starship *Enterprise* went through a wormhole, something unexpected happened. I know, because I was on board, playing poker with Newton, Einstein, and Data. I had a big surprise. Just look what appeared on my knee.

CHAPTER 5

PROTECTING THE PAST

Is time travel possible?
Could an advanced civilization go back and change the past?

Whereas Stephen W. Hawking (having lost a previous bet on this subject by not demanding genericity) still firmly believes that naked singularities are an anathema and should be prohibited by the laws of classical physics,

And whereas John Preskill and Kip Thorne (having won the previous bet) still regard naked singularities as quantum gravitational objects that might exist, unclothed by horizons, for all the Universe to see,

Therefore Hawking offers, and Preskill/Thorne accept, a wager that

> When any form of classical matter or field that is incapable of becoming singular in flat spacetime is coupled to general relativity via the classical Einstein equations, then

A dynamical evolution from generic initial conditions (*i.e., from an open set of initial data*) **can never produce a naked singularity** (*a past-incomplete null geodesic from $\mathcal{I}_+$*).

The loser will reward the winner with clothing to cover the winner's nakedness. The clothing is to be embroidered with a suitable, truly concessionary message.

Stephen W. Hawking   John P. Preskill & Kip S. Thorne

Pasadena, California, 5 February 1997

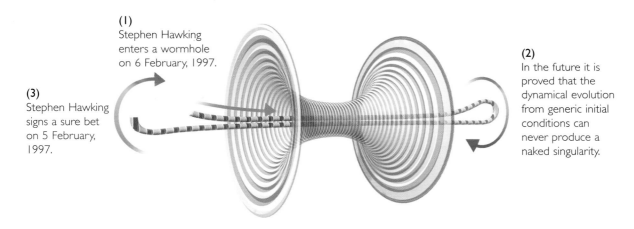

(1)
Stephen Hawking
enters a wormhole
on 6 February, 1997.

(3)
Stephen Hawking
signs a sure bet
on 5 February,
1997.

(2)
In the future it is
proved that the
dynamical evolution
from generic initial
conditions can
never produce a
naked singularity.

M Y FRIEND AND COLLEAGUE KIP THORNE, WITH WHOM
I have had a number of bets (left), is not one to follow
the accepted line in physics just because everyone else
does. This led him to have the courage to be the first serious scientist to discuss time travel as a practical possibility.

It is tricky to speculate openly about time travel. One risks
either an outcry at the waste of public money being spent on something so ridiculous or a demand that the research be classified for
military purposes. After all, how could we protect ourselves against
someone with a time machine? They might change history and rule
the world. There are only a few of us foolhardy enough to work on
a subject that is so politically incorrect in physics circles. We dis-
guise the fact by using technical terms that are code for time travel.

Kip Thorne

133

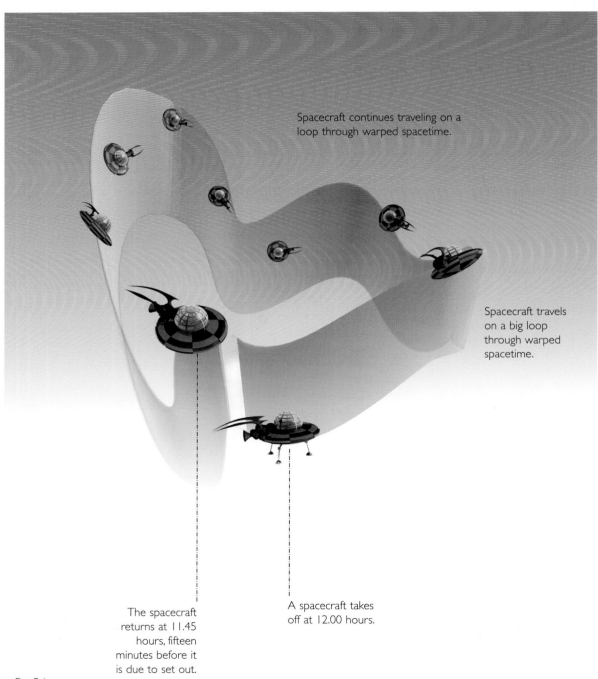

Spacecraft continues traveling on a loop through warped spacetime.

Spacecraft travels on a big loop through warped spacetime.

The spacecraft returns at 11.45 hours, fifteen minutes before it is due to set out.

A spacecraft takes off at 12.00 hours.

FIG. 5.1

The basis of all modern discussions of time travel is Einstein's general theory of relativity. As we have seen in earlier chapters, the Einstein equations made space and time dynamic by describing how they were curved and distorted by the matter and energy in the universe. In general relativity someone's personal time as measured by their wristwatch would always increase, just as it did in Newtonian theory or the flat spacetime of special relativity. But there was now the possibility that spacetime could be warped so much that you could go off in a spaceship and come back before you set out (Fig. 5.1).

One way this could happen is if there were wormholes, tubes of spacetime mentioned in Chapter 4 that connect different regions of space and time. The idea is that you steer your spaceship into one mouth of the wormhole and come out of the other mouth in a different place and at a different time (Fig. 5.2, see page 136).

Wormholes, if they exist, would be the solution to the speed limit problem in space: it would take tens of thousands of years to cross the galaxy in a spaceship that traveled at less than the speed of light, as relativity demands. But you might go through a wormhole to the other side of the galaxy and be back in time for dinner. However, one can show that if wormholes exist, you could also use them to get back before you set out. So you might think that you could do something like blowing up the rocket on its launch pad to prevent your setting out in the first place. This is a variation of the grandfather paradox: what happens if you go back and kill your grandfather before your father was conceived? (see Fig. 5.3, page 138)

Of course, this is a paradox only if you believe you have free will to do what you like when you go back in time. This book will

SHALLOW WORMHOLE

Enters at 12.00 hours   Exits at 12.00 hours

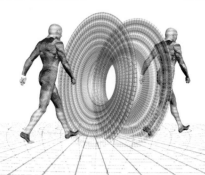

(FIG. 5.2) A SECOND VARIATION ON THE TWINS PARADOX

(1)
If there was a wormhole with its two ends close together,
you could walk through the wormhole and come out at the same time.

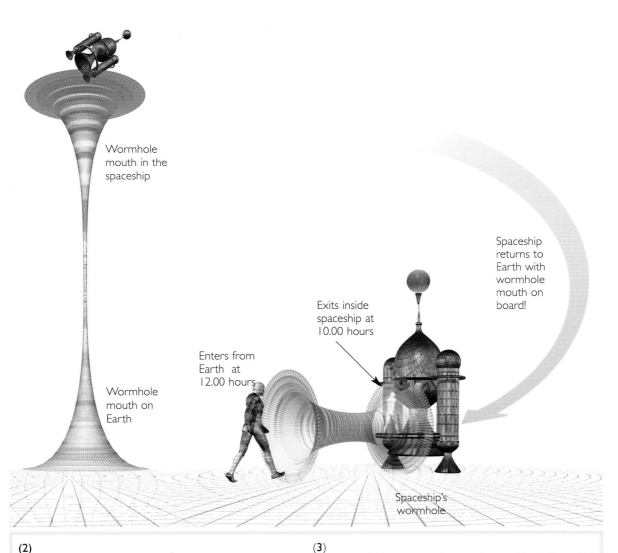

Wormhole mouth in the spaceship

Spaceship returns to Earth with wormhole mouth on board!

Exits inside spaceship at 10.00 hours

Enters from Earth at 12.00 hours

Wormhole mouth on Earth

Spaceship's wormhole

(2)
One can imagine taking one end of the wormhole on a long journey on a spaceship while the other end remains on Earth.

(3)
Because of the twins paradox effect, when the spaceship returns, less time has elapsed for the mouth it contains than for the mouth that stays on earth. This would mean that if you step into the Earth mouth, you could come out of the spaceship at an earlier time.

(FIG. 5.3)
Can a bullet fired through a wormhole into an earlier time affect the one who fires it?

COSMIC STRINGS

Cosmic strings are long, heavy objects with a tiny cross section that may have been produced during the early stages of the universe. Once cosmic strings formed, they were further stretched by the expansion of the universe, and by now a single cosmic string could cross over the entire length of our observable universe.

The occurrence of cosmic strings is suggested by modern theories of particles, which predict that in the hot early stages of the universe, matter was in a symmetric phase, much like liquid water—which is symmetrical: the same at every point in every direction—rather than like ice crystals, which have a discrete structure.

When the universe cooled, the symmetry of the early phase could have been broken in different ways in distant regions. Consequently, the cosmic matter would have settled into different ground states in those regions. Cosmic strings are the configurations of matter at the boundaries between these regions. Their formation was therefore an inevitable consequence of the fact that different regions could not agree on their ground states.

not go into a philosophical discussion of free will. Instead it will concentrate on whether the laws of physics allow spacetime to be so warped that a macroscopic body such as a spaceship can return to its own past. According to Einstein's theory, a spaceship necessarily travels at less than the local speed of light and follows what is called a timelike path through spacetime. Thus one can formulate the question in technical terms: does spacetime admit timelike curves that are closed—that is, that return to their starting point again and again? I shall refer to such paths as "time loops."

There are three levels on which we can try to answer this question. The first is Einstein's general theory of relativity, which assumes that the universe has a well-defined history without any uncertainty. For this classical theory we have a fairly complete picture. However, as we have seen, this theory can't be quite right, because we observe that matter is subject to uncertainty and quantum fluctuations.

We can therefore ask the question about time travel on a second level, that of semiclassical theory. In this, we consider matter to behave according to quantum theory, with uncertainty and quantum fluctuations, but spacetime to be well defined and classical. Here the picture is less complete, but at least we have some idea of how to proceed.

(Fig. 5.4)
Does spacetime admit timelike curves that are closed, returning to their starting point again and again?

Finally, there is the full quantum theory of gravity, whatever that may be. In this theory, where not just matter but also time and space themselves are uncertain and fluctuate, it is not even clear how to pose the question of whether time travel is possible. Maybe the best we can do is to ask how people in regions where spacetime is nearly classical and free from uncertainty would interpret their measurements. Would they think that time travel had taken place in regions of strong gravity and large quantum fluctuations?

To start with the classical theory: the flat spacetime of special relativity (relativity without gravity) doesn't allow time travel, nor do the curved spacetimes that were known early on. It was therefore a great shock to Einstein when in 1949 Kurt Gödel, of Gödel's theorem (see box), discovered a spacetime that was a universe full of rotating matter, with time loops through every point (Fig. 5.4).

The Gödel solution required a cosmological constant, which may or may not exist in nature, but other solutions were subsequently found without a cosmological constant. A particularly interesting case is one in which two cosmic strings move at high speed past each other.

Cosmic strings should not be confused with the strings of

GÖDEL'S INCOMPLETENESS THEOREM

In 1931 the mathematician Kurt Gödel proved his famous incompleteness theorem about the nature of mathematics. The theorem states that within any formal system of axioms, such as present-day mathematics, questions always persist that can neither be proved nor disproved on the basis of the axioms that define the system. In other words, Gödel showed that there are problems that cannot be solved by any set of rules or procedures.

Gödel's theorem set fundamental limits on mathematics. It came as a great shock to the scientific community, since it overthrew the widespread belief that mathematics was a coherent and complete system based on a single logical foundation. Gödel's theorem, Heisenberg's uncertainty principle, and the practical impossibility of following the evolution of even a deterministic system that becomes chaotic, form a core set of limitations to scientific knowledge that only came to be appreciated during the twentieth century.

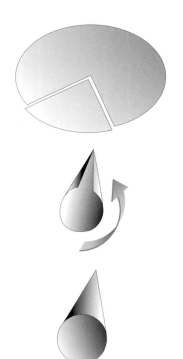

FIG. 5.5

string theory, though they are not entirely unrelated. They are objects with length but a tiny cross section. Their occurrence is predicted in some theories of elementary particles. The spacetime outside a single cosmic string is flat. However, it is flat spacetime with a wedge cut out, with the sharp end of the wedge at the string. It is like a cone: take a large circle of paper and cut out a segment like a slice of pie, a wedge with its corner at the center of the circle. Then discard the piece you have cut out and glue the cut edges of the remaining piece together so that you have a cone. This represents the spacetime in which the cosmic string exists (Fig. 5.5).

Notice that because the surface of the cone is the same flat sheet of paper with which you started (minus the wedge), you can still call it "flat" except at the apex. You can recognize that there is curvature at the apex by the fact that a circle around the apex has a smaller circumference than a circle drawn at the same distance around the center of the original round sheet of paper. In other words, a circle around the apex is shorter than one would expect for a circle of that radius in flat space because of the missing segment (Fig. 5.6).

Similarly, in the case of a cosmic string, the wedge that is removed from flat spacetime shortens circles around the string but does not affect time or distances along the string. This means that the spacetime around a single cosmic string does not contain any time loops, so it is not possible to travel into the past. However, if there is a second cosmic string that is moving relative to the first, its time direction will be a combination of the time and space directions of the first. This means that the wedge that is cut out for the second string will shorten both distances in space and time intervals as seen by someone moving with the first string (Fig. 5.7). If the cosmic strings are moving at nearly the speed of light relative to each other, the saving of time going around both strings can be so great that one arrives back before one set out. In other words, there are time loops that one can follow to travel into the past.

The cosmic string spacetime contains matter that has positive energy density and is consistent with the physics we know. However, the warping that produces the time loops extends all the way out to infinity in space and back to the infinite past in time. Thus these

FIG. 5.6

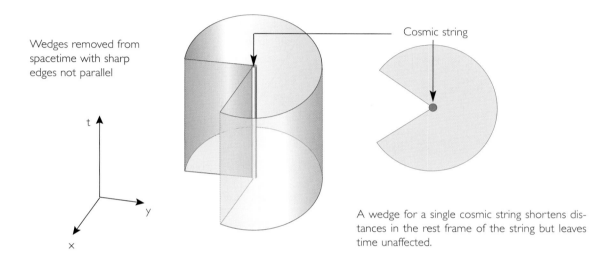

Wedges removed from spacetime with sharp edges not parallel

Cosmic string

A wedge for a single cosmic string shortens distances in the rest frame of the string but leaves time unaffected.

FIG. 5.7

A second wedge cut out for another moving cosmic string will shorten distances in both space and time in the rest frame of the first cosmic string.

FINITELY GENERATED TIME TRAVEL HORIZON

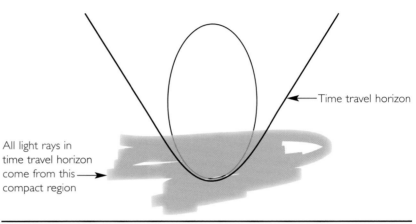

Time travel horizon

All light rays in
time travel horizon
come from this
compact region

S

(FIG. 5.8)
Even the most advanced civilization could warp spacetime only in a finite region. The time travel horizon, the boundary of the part of spacetime in which it is possible to travel into one's past, would be formed by light rays that emerge from finite regions.

spacetimes were created with time travel in them. We have no reason to believe that our own universe was created in such a warped fashion, and we have no reliable evidence of visitors from the future. (I'm discounting the conspiracy theory that UFOs are from the future and that the government knows and is covering it up. Its record of cover-ups is not that good.) I shall therefore assume that there were no time loops in the distant past or, more precisely, in the past of some surface through spacetime that I shall call S. The question then is: could some advanced civilization build a time machine? That is, could it modify the spacetime to the future of S (above the surface S in the diagram) so that time loops appeared in a finite region? I say a finite region because no matter how advanced the civilization becomes, it could presumably control only a finite part of the universe.

In science, finding the right formulation of a problem is often the key to solving it, and this was a good example. To define what was meant by a finite time machine, I went back to some early work of mine. Time travel is possible in a region of spacetime in which there are time loops, paths that move at less than the speed of light but which nevertheless manage to come back to the place and time they started because of the warping of spacetime. Since I have

assumed there were no time loops in the distant past, there must be what I call a time travel "horizon," the boundary separating the region of time loops from the region without them (Fig.5.8).

Time travel horizons are rather like black hole horizons. While a black hole horizon is formed by light rays that just miss falling into the black hole, a time travel horizon is formed by light rays on the verge of meeting up with themselves. I then take as my criterion for a time machine what I call a finitely generated horizon—that is, a horizon that is formed by light rays that all emerge from a bounded region. In other words, they don't come in from infinity or from a singularity, but originate from a finite region containing time loops—the sort of region our advanced civilization is supposed to create.

In adopting this definition as the footprint of a time machine, we have the advantage of being able to use the machinery that Roger Penrose and I developed to study singularities and black holes. Even without using the Einstein equations, I can show that, in general, a finitely generated horizon will contain a light ray that actually meets up with itself—that is, a light ray that keeps coming back to the same point over and over again. Each time the light came around it would be more and more blue-shifted, so the images would get bluer and bluer. The wave crests of a pulse of light will get closer and closer together and the light will get around in shorter and shorter intervals of its time. In fact, a particle of light would have only a finite history, as defined by its own measure of time, even though it went around and around in a finite region and did not hit a curvature singularity.

The question then is: could some advanced civilization build a time machine?

143

(FIG. 5.9, above)
The danger of time travel.

(FIG. 5.10, opposite)
The prediction that black holes radiate and lose mass implies that quantum theory causes negative energy to flow into the black hole across the horizon. For the black hole to shrink in size, the energy density on the horizon must be negative, the sign that is required to build a time machine.

One might not care if a particle of light completed its history in a finite time. But I can also prove that there would be paths moving at less than the speed of light that had only finite duration. These could be the histories of observers who would be trapped in a finite region before the horizon and would go around and around faster and faster until they reached the speed of light in a finite time. So if a beautiful alien in a flying saucer invites you into her time machine, step with care. You might fall into one of these trapped repeating histories of only finite duration (Fig. 5.9).

These results do not depend on the Einstein equations but only on the way spacetime would have to warp to produce time loops in a finite region. However, we can now ask what kind of matter an advanced civilization would have to use to warp spacetime so as to build a finite-sized time machine. Can it have positive energy density everywhere, as in the cosmic string spacetime I described earlier? The cosmic string spacetime did not satisfy my requirement that the time loops appear in a finite region. However, one might think that this was just because the cosmic strings were infinitely long. One might imagine that one could build a finite time machine using finite loops of cosmic string and have the energy density positive everywhere. It is a pity to disappoint people such as Kip, who want to return to the past, but it can't be done with positive energy density everywhere. I can prove that to build a finite time machine, you need negative energy.

Energy density is always positive in classical theory, so time machines of finite size are ruled out on this level. However, the situation is different in the semiclassical theory, in which one considers matter to behave according to quantum theory but spacetime to be well defined and classical. As we have seen, the uncertainty principle of quantum theory means that fields are always fluctuating up and down even in apparently empty space, and have an energy density that is infinite. Thus one has to subtract an infinite quantity to get the finite energy density that we observe in the universe. This subtraction can leave the energy density negative, at least locally. Even in flat space, one can find quantum states in which the energy

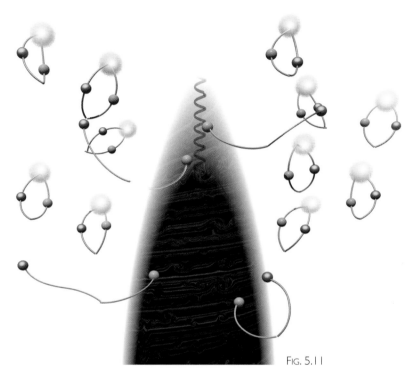

FIG. 5.11

density is negative locally although the total energy is positive. One might wonder whether these negative values actually cause spacetime to warp in the appropriate way to build a finite time machine, but it seems they must. As we saw in Chapter 4, quantum fluctuations mean that even apparently empty space is full of pairs of virtual particles that appear together, move apart, and then come back together and annihilate each other (Fig. 5.10). One member of a virtual-particle pair will have positive energy and the other negative energy. When a black hole is present, the negative-energy member can fall in and the positive-energy member can escape to infinity, where it appears as radiation that carries positive energy away from the black hole. The negative-energy particles falling in cause the black hole to lose mass and to evaporate slowly, with its horizon shrinking in size (Fig. 5.11).

Ordinary matter with positive energy density has an attractive gravitational effect and warps spacetime to bend light rays toward each other—just as the ball on the rubber sheet in Chapter 2 always makes the smaller ball bearings curve toward it, never away.

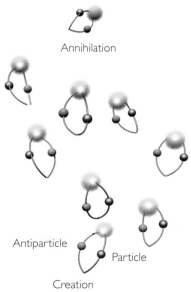

Annihilation

Antiparticle

Particle

Creation

FIG. 5.10

145

My grandson,
William Mackenzie Smith.

This would imply that the area of the horizon of a black hole could only increase with time, never shrink. For the horizon of a black hole to shrink in size, the energy density on the horizon must be negative and warp spacetime to make light rays diverge from each other. This was something I first realized when I was getting into bed soon after the birth of my daughter. I won't say how long ago that was, but I now have a grandson.

The evaporation of black holes shows that on the quantum level the energy density can sometimes be negative and warp spacetime in the direction that would be needed to build a time machine. Thus we might imagine that some very advanced civilization could arrange things so that the energy density is sufficiently negative to form a time machine that could be used by macroscopic objects such as spaceships. However, there's an important difference between a black hole horizon, which is formed by light rays that just keep going, and the horizon in a time machine, which contains closed light rays that keep going around and around. A virtual particle moving on such a closed path would bring its ground state energy back to the same point again and again. One would therefore expect the energy density to be infinite on the horizon—the boundary of the time machine, the region in which one can travel into the past. This is borne out by explicit calculations in a few backgrounds that are simple enough for exact calculations. It would mean that a person or a space probe that tried to cross the horizon to get into the time machine would get wiped out by a bolt of radiation (Fig. 5.12). So the future looks black for time travel—or should one say blindingly white?

The energy density of matter depends on the state it is in, so it is possible that an advanced civilization might be able to make the energy density finite on the boundary of the time machine by "freezing out" or removing the virtual particles that go around and around in a closed loop. It is not clear, however, that such a time machine would be stable: the least disturbance, such as someone crossing the horizon to enter the time machine, might set off circulating virtual particles and trigger a bolt of lightning. This is a

(Fig. 5.12)
One might get wiped out by a bolt
of radiation when crossing the time
travel horizon.

question that physicists should be free to discuss without being laughed to scorn. Even if it turns out that time travel is impossible, it is important that we understand why it is impossible.

To answer that question definitively, we need to consider quantum fluctuations not only of matter fields but of spacetime itself. One might expect that these would cause a certain fuzziness in the paths of light rays and in the whole concept of time ordering. Indeed, one can regard the radiation from black holes as leaking out because quantum fluctuations of spacetime mean that the horizon is not exactly defined. Because we don't yet have a complete theory of quantum gravity, it is difficult to say what the effects of spacetime fluctuations should be. Nevertheless, we can hope to get some pointers from the Feynman sum over histories described in Chapter 3.

Closed loop

Histories of particle

Path of the
particle traveling
backward in time
on a loop

Particle

TIME

SPACE

SUM OVER PARTICLE HISTORIES

(FIG. 5.13)

The Feynman sum over histories has
to include histories in which particles
travel back in time, and even histories
that are closed loops in time and
space.

Each history will be a curved spacetime with matter fields in it.
Since we are supposed to sum over all possible histories, not just those
that satisfy some equations, the sum must include spacetimes that are
warped enough for travel into the past (Fig. 5.13). So the question is,
why isn't time travel happening everywhere? The answer is that time
travel is indeed taking place on a microscopic scale, but we don't
notice it. If one applies the Feynman sum-over-histories idea to a particle, one has to include histories in which the particle travels faster
than light and even backward in time. In particular, there will be histories in which the particle goes around and around on a closed loop
in time and space. It would be like the film *Groundhog Day*, in which
a reporter has to live the same day over and over again (Fig. 5.14).

One cannot observe particles with such closed-loop histories
directly with a particle detector. However, their indirect effects have
been measured in a number of experiments. One is a small shift in the

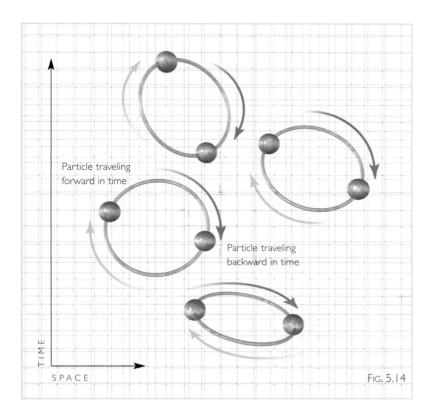

Particle traveling forward in time

Particle traveling backward in time

TIME

SPACE

FIG. 5.14

light given out by hydrogen atoms, caused by electrons moving in closed loops. Another is a small force between parallel metal plates, caused by the fact that there are slightly fewer closed-loop histories that can fit between the plates compared to the region outside— another equivalent interpretation of the Casimir effect. Thus the existence of closed-loop histories is confirmed by experiment (Fig. 5.15).

One might dispute whether closed-loop particle histories have anything to do with the warping of spacetime, because they occur even in fixed backgrounds such as flat space. But in recent years we have found that phenomena in physics often have dual, equally valid descriptions. One can equally well say that a particle moves on a closed loop in a given fixed background, or that the particle stays fixed and space and time fluctuate around it. It is just a question of whether you do the sum over particle paths first and then the sum over curved spacetimes, or vice versa.

FIG. 5.15

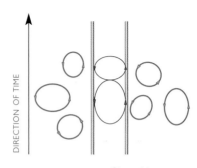

DIRECTION OF TIME

Closed loops

149

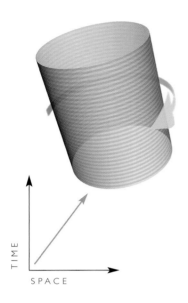

TIME

SPACE

(FIG. 5.16)
The Einstein universe is like a cylinder: it is finite in space and constant in time. Because of its finite size, it can rotate at less than the speed of light everywhere.

It seems, therefore, that quantum theory allows time travel on a microscopic scale. However, this is not much use for science fiction purposes, such as going back and killing your grandfather. The question therefore is: can the probability in the sum over histories be peaked around spacetimes with macroscopic time loops?

One can investigate this question by studying the sum over histories of matter fields in a series of background spacetimes that get closer and closer to admitting time loops. One would expect something dramatic to happen when time loops first appear, and this is borne out in a simple example that I studied with my student Michael Cassidy.

The background spacetimes in the series we studied were closely related to what is called the Einstein universe, the spacetime that Einstein proposed when he believed that the universe was static and unchanging in time, neither expanding nor contracting (see Chapter 1). In the Einstein universe time runs from the infinite past to the infinite future. The space directions, however, are finite and close on themselves, like the surface of the Earth but with one more dimension. One can picture this spacetime as a cylinder with the long axis being the time direction and the cross section being the three space directions (Fig. 5.16).

The Einstein universe does not represent the universe we live in because it is not expanding. Nevertheless, it is a convenient background to use when discussing time travel, because it is simple enough that one can do the sum over histories. Forgetting about time travel for the moment, consider matter in an Einstein universe that is rotating about some axis. If you were on the axis, you could remain at the same point of space, just as you do when standing at the center of a children's carousel. But if you were not on the axis, you would be moving through space as you rotated about the axis. The further you were from the axis, the faster you would be moving (Fig. 5.17). So if the universe were infinite in space, points sufficiently far from the axis would have to be rotating faster than light. However, because the Einstein universe is finite in the space directions, there is a critical rate of rotation below which no part of the universe is rotating faster than light.

ROTATING IN FLAT SPACE

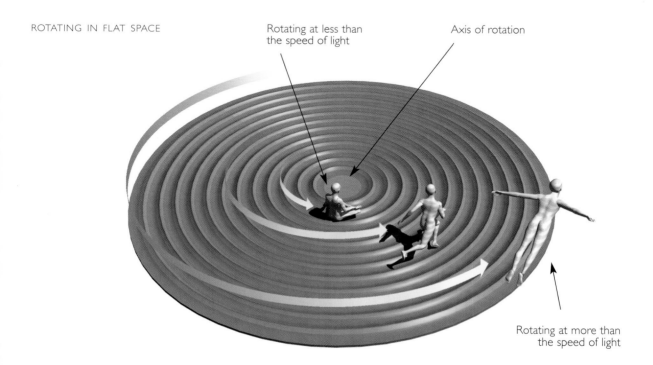

Rotating at less than the speed of light

Axis of rotation

Rotating at more than the speed of light

Now consider the sum over particle histories in a rotating Einstein universe. When the rotation is slow, there are many paths a particle can take using a given amount of energy. Thus the sum over all particle histories in this background gives a large amplitude. This means that the probability of this background would be high in the sum over all curved spacetime histories—that is, it is among the more probable histories. However, as the rate of rotation of the Einstein universe approaches the critical value, so that its outer edges are moving at a speed approaching the speed of light, there is only one particle path that is classically allowed on that edge, namely, one that is moving at the speed of light. This means that the sum over particle histories will be small. Thus the probability of these backgrounds will be low in the sum over all curved spacetime histories. That is, they are the least probable.

(Fig. 5.17)
In flat space a rigid rotation will move faster than the speed of light far from its axis.

151

(FIG. 5.18) BACKGROUND WITH CLOSED TIME-LIKE CURVES

Universe expanding in this direction

Universe *not* expanding in this direction

Identify with a boost in vertical speed

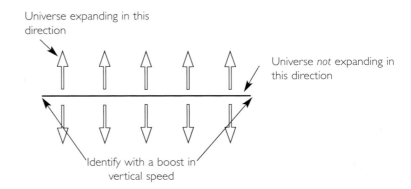

What do rotating Einstein universes have to do with time travel and time loops? The answer is that they are mathematically equivalent to other backgrounds that do admit time loops. These other backgrounds are universes that are expanding in two space directions. The universes are not expanding in the third space direction, which is periodic. That is to say, if you go a certain distance in this direction, you get back to where you started. However, each time you do a circuit of the third space direction, your speed in the first or second directions is increased (Fig. 5.18).

If the boost is small, there are no time loops. However, consider a sequence of backgrounds with increasing boosts in speed. At a certain critical boost, time loops will appear. Not surprisingly, this critical boost corresponds to the critical rate of rotation of the Einstein universes. Since the sum-over-histories calculations in these backgrounds are mathematically equivalent, one can conclude that the probability of these backgrounds goes to zero as they approach the warping needed for time loops. In other words, the probability of having sufficient warping for a time machine is zero. This supports what I have called the Chronology Protection Conjecture: that the laws of physics conspire to prevent time travel by macroscopic objects.

Although time loops are allowed by the sum over histories, the probabilities are extremely small. Based on the duality arguments I mentioned earlier, I estimate the probability that Kip Thorne could go back and kill his grandfather as less than one in ten with a trillion trillion trillion trillion trillion zeroes after it.

That's a pretty small probability, but if you look closely at the picture of Kip, you may see a slight fuzziness around the edges. That corresponds to the faint possibility that some bastard from the future came back and killed his grandfather, so he's not really there.

As gambling men, Kip and I would bet on odds like that. The trouble is, we can't bet each other because we are now both on the same side. On the other hand, I wouldn't take a bet with anyone else. He might be from the future and know that time travel worked.

You might wonder if this chapter is part of a government cover-up on time travel. You might be right.

The probability that Kip could go back and kill his grandfather is $1/10^{10^{60}}$.

In other words less than 1 in 10 —with a trillion trillion trillion trillion trillion zeroes after it.

CHAPTER 6

OUR FUTURE? STAR TREK OR NOT?

*How biological and electronic life
will go on developing in complexity at an ever-increasing rate.*

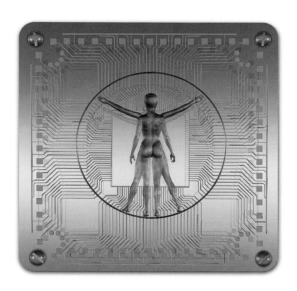

(FIG. 6.1) GROWTH OF POPULATION

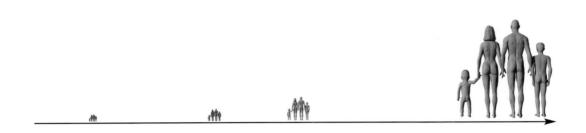

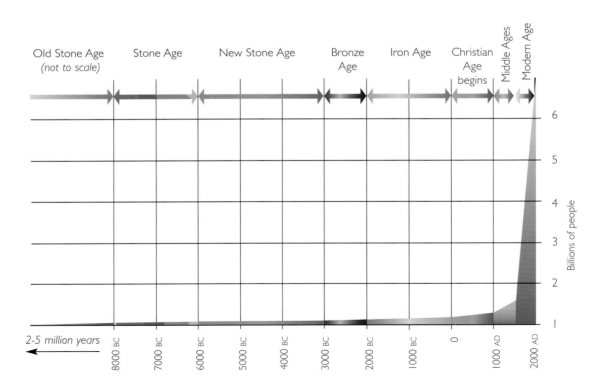

Newton, Einstein, Commander Data, and myself playing poker in a scene from Star Trek.

THE REASON *STAR TREK* IS SO POPULAR IS BECAUSE IT IS A safe and comforting vision of the future. I'm a bit of a *Star Trek* fan myself, so I was easily persuaded to take part in an episode in which I played poker with Newton, Einstein, and Commander Data. I beat them all, but unfortunately there was a red alert, so I never collected my winnings.

Star Trek shows a society that is far in advance of ours in science, in technology, and in political organization. (The last might not be difficult.) There must have been great changes, with their accompanying tensions and upsets, in the time between now and then, but in the period we are shown, science, technology, and the organization of society are supposed to have achieved a level of near perfection.

I want to question this picture and ask if we will ever reach a final steady state in science and technology. At no time in the ten thousand years or so since the last ice age has the human race been in a state of constant knowledge and fixed technology. There have been a few setbacks, like the Dark Ages after the fall of the Roman Empire. But the world's population, which is a measure of our technological ability to preserve life and feed ourselves, has risen steadily, with only a few hiccups such as the Black Death (Fig. 6.1).

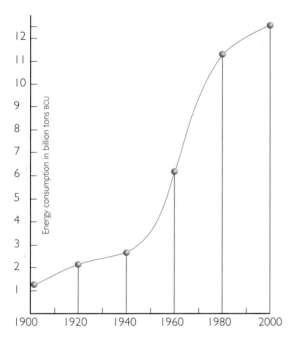

WORLDWIDE ELECTRICITY CONSUMPTION

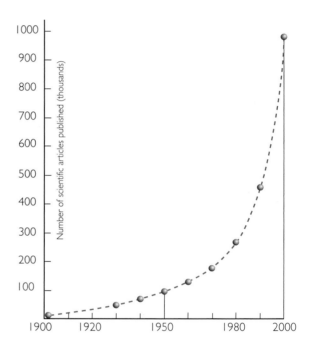

WORLDWIDE PUBLICATION OF SCIENTIFIC ARTICLES

(FIG 6.2)

Left: The total worldwide energy consumption in billions of tons BCU, where 1 ton ~Bituminous Coal Unit = 8.13 MW-hr.

Right: The number of scientific articles published each year. The vertical scale is in thousands. In 1900 there were 9,000. By 1950 there were 90,000 and by the year 2000 there were 900,000.

In the last two hundred years, population growth has become exponential; that is, the population grows by the same percentage each year. Currently, the rate is about 1.9 percent a year. That may not sound like very much, but it means that the world population doubles every forty years (Fig. 6.2).

Other measures of technological development in recent times are electricity consumption and the number of scientific articles. They too show exponential growth, with doubling times of less than forty years. There is no sign that scientific and technological development will slow down and stop in the near future—certainly not by the time of *Star Trek*, which is supposed to be not that far in the future. But if the population growth and the increase in the consumption of electricity continue at their current rates, by 2600 the world's population will be standing shoulder to shoulder, and electricity use will make the Earth glow red-hot (see illustration opposite).

By the year 2600 the world's population would be standing shoulder to shoulder, and the electricity consumption would make the Earth glow red-hot.

If you stacked all the new books being published next to each other, you would have to move at ninety miles an hour just to keep up with the end of the line. Of course, by 2600 new artistic and scientific work will come in electronic forms, rather than as physical books and papers. Nevertheless, if the exponential growth continued, there would be ten papers a second in my kind of theoretical physics, and no time to read them.

Clearly, the present exponential growth cannot continue indefinitely. So what will happen? One possibility is that we will wipe ourselves out completely by some disaster, such as a nuclear war. There is a sick joke that the reason we have not been contacted by extraterrestrials is that when a civilization reaches our stage of development, it becomes unstable and destroys itself. However, I'm an optimist. I don't believe the human race has come so far just to snuff itself out when things are getting interesting.

(FIG. 6.3)

Star Trek's story line depends on the *Enterprise*, and starships like the one above, being able to travel at warp speed, which is much faster than light. However, if the Chronology Protection Conjecture is correct, we shall have to explore the galaxy using rocket-propelled spaceships that travel slower than light.

The *Star Trek* vision of the future—that we achieve an advanced but essentially static level—may come true in respect of our knowledge of the basic laws that govern the universe. As I shall describe in the next chapter, there may be an ultimate theory that we will discover in the not-too-distant future. This ultimate theory, if it exists, will determine whether the *Star Trek* dream of warp drive can be realized. According to present ideas, we shall have to explore the galaxy in a slow and tedious manner, using spaceships traveling slower than light, but since we don't yet have a complete unified theory, we can't quite rule out warp drive (Fig. 6.3).

On the other hand, we already know the laws that hold in all but the most extreme situations: the laws that govern the crew of the *Enterprise,* if not the spaceship itself. Yet it doesn't seem that we will ever reach a steady state in the uses we make of these laws or in the complexity of the systems that we can produce with them. It is with this complexity that the rest of this chapter will be concerned.

By far the most complex systems that we have are our own bodies. Life seems to have originated in the primordial oceans that covered the Earth four billion years ago. How this happened we don't know. It may be that random collisions between atoms built up macromolecules that could reproduce themselves and assemble themselves into more complicated structures. What we do know is that by three and a half billion years ago, the highly complicated DNA molecule had emerged.

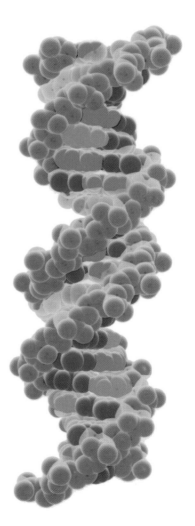

DNA is the basis for all life on Earth. It has a double helix structure, like a spiral staircase, which was discovered by Francis Crick and James Watson in the Cavendish lab at Cambridge in 1953. The two strands of the double helix are linked by pairs of bases, like the treads in a spiral staircase. There are four bases in DNA: adenine, guanine, thymine, and cytosine. The order in which they occur along the spiral staircase carries the genetic information that enables the DNA to assemble an organism around it and reproduce itself. As it makes copies of itself, there are occasional errors in the proportion or order of the bases along the spiral. In most cases, the mistakes in copying make the DNA either unable or less likely to reproduce itself, meaning that such genetic errors, or mutations, as they are called, will die out. But in a few cases, the error or mutation will increase the chances of the DNA surviving and reproducing. Such changes in the genetic code will be favored. This is how the information contained in the sequence of DNA gradually evolves and increases in complexity (see Fig. 6.4, page 162).

Because biological evolution is basically a random walk in the

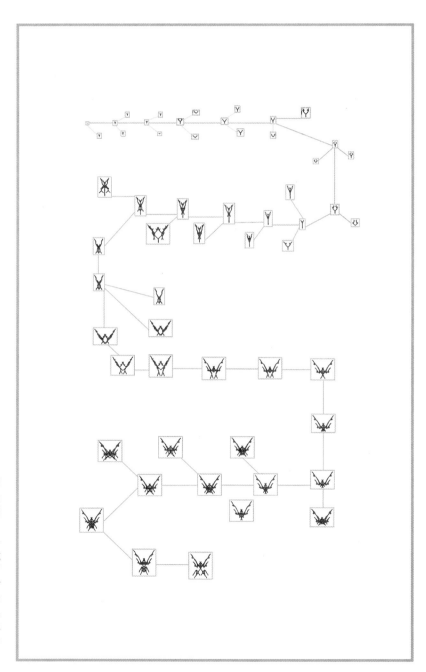

(FIG. 6.4) EVOLUTION IN ACTION

On the right are computer-generated biomorphs that evolved in a program devised by the biologist Richard Dawkins.

Survival of a particular strain depended upon simple qualities like being "interesting," "different," or "insect-like."

Starting from a single pixel, the early random generations developed through a process similar to natural selection. Dawkins bred an insect-like form in a remarkable 29 generations (with a number of evolutionary dead ends).

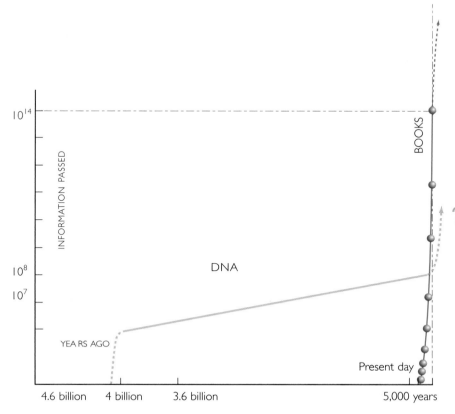

BOOKS

The development of
complexity since the
formation of the
Earth (not to scale).

10^{14}

INFORMATION PASSED

DNA

10^8

10^7

YEARS AGO

Present day

?

4.6 billion   4 billion   3.6 billion   5,000 years

space of all genetic possibilities, it has been very slow. The complexity, or number of bits of information, that is coded in DNA is roughly the number of bases in the molecule. For the first two billion years or so, the rate of increase in complexity must have been of the order of one bit of information every hundred years. The rate of increase of DNA complexity gradually rose to about one bit a year over the last few million years. But then, about six or eight thousand years ago, a major new development occurred. We developed written language. This meant that information could be passed from one generation to the next without having to wait for the very slow process of random mutations and natural selection to code it into the DNA sequence. The amount of complexity increased enormously. A single paperback romance could hold as much information as the difference in DNA between apes and humans, and a thirty-volume encyclopedia could describe the entire sequence of human DNA (Fig. 6.5).

Even more important, the information in books can be

← Entire human DNA sequence in 30 volumes →

FIG 6.5

163

Growing embryos outside the human body will allow bigger brains and greater intelligence.

updated rapidly. The current rate at which human DNA is being updated by biological evolution is about one bit a year. But there are two hundred thousand new books published each year, a new-information rate of over a million bits a second. Of course, most of this information is garbage, but even if only one bit in a million is useful, that is still a hundred thousand times faster than biological evolution.

This transmission of data through external, nonbiological means has led the human race to dominate the world and to have an exponentially increasing population. But now we are at the beginning of a new era, in which we will be able to increase the complexity of our internal record, the DNA, without having to wait for the slow process of biological evolution. There has been no significant change in human DNA in the last ten thousand years, but it is likely that we will be able to completely redesign it in the next thousand. Of course, many people will say that genetic engineering of humans should be banned, but it is doubtful we will be able to prevent it. Genetic engineering of plants and animals will be allowed for economic reasons, and someone is bound to try it on humans. Unless we have a totalitarian world order, someone somewhere will design improved humans.

Clearly, creating improved humans will create great social and political problems with respect to unimproved humans. My intention is not to defend human genetic engineering as a desirable development, but just to say it is likely to happen whether we want it or not. This is the reason why I don't believe science fiction like *Star Trek*, where people four hundred years into the future are essentially the same as we are today. I think the human race, and its DNA, will increase its complexity quite rapidly. We should recognize that this is likely to happen and consider how we will deal with it.

In a way, the human race needs to improve its mental and physical qualities if it is to deal with the increasingly complex world around it and meet new challenges such as space travel. Humans also need to increase their complexity if biological systems are to keep ahead of electronic ones. At the moment, computers have the advantage of speed, but they show no sign of intelligence. This is not surprising, because our present computers are less complex than the brain of an earthworm, a species not noted for its intellectual powers.

But computers obey what is known as Moore's law: their speed

At present our computers remain outstripped in computational power by the brain of a humble earthworm.

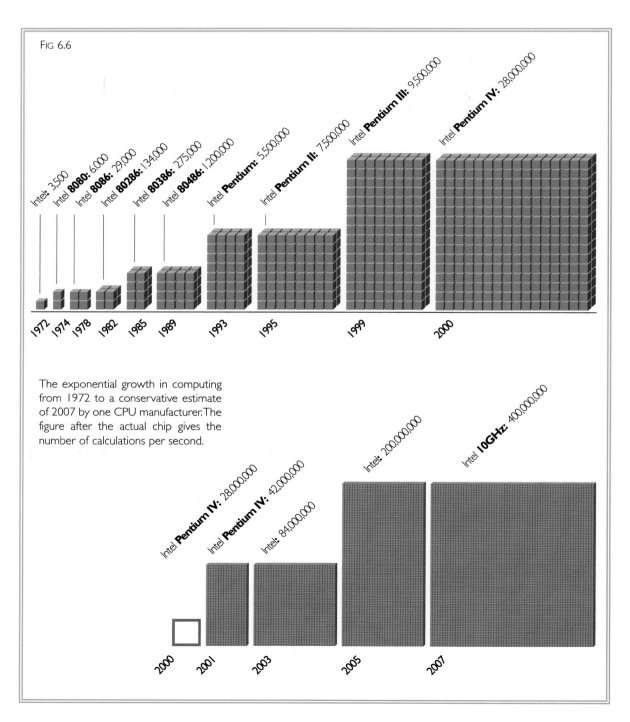

FIG 6.6

Intel: 3,500
Intel **8080**: 6,000
Intel **8086**: 29,000
Intel **80286**: 134,000
Intel **80386**: 275,000
Intel **80486**: 1,200,000
Intel **Pentium**: 5,500,000
Intel **Pentium II**: 7,500,000
Intel **Pentium III**: 9,500,000
Intel **Pentium IV**: 28,000,000

1972 1974 1978 1982 1985 1989 1993 1995 1999 2000

The exponential growth in computing from 1972 to a conservative estimate of 2007 by one CPU manufacturer. The figure after the actual chip gives the number of calculations per second.

Intel **Pentium IV**: 28,000,000
Intel **Pentium IV**: 42,000,000
Intel: 84,000,000
Intel: 200,000,000
Intel **10GHz**: 400,000,000

2000 2001 2003 2005 2007

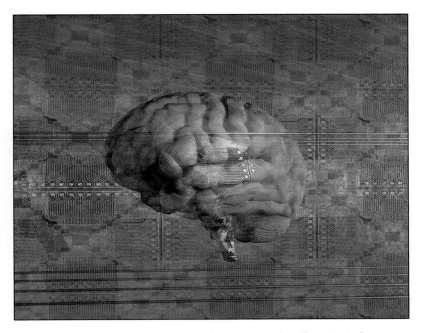

and complexity double every eighteen months (Fig. 6.6). It is one of those exponential growths that clearly cannot continue indefinitely. However, it will probably continue until computers have a complexity similar to that of the human brain. Some people say that computers can never show true intelligence, whatever that may be. But it seems to me that if very complicated chemical molecules can operate in humans to make them intelligent, then equally complicated electronic circuits can also make computers act in an intelligent way. And if they are intelligent, they can presumably design computers that have even greater complexity and intelligence.

Will this increase of biological and electronic complexity go on forever, or is there a natural limit? On the biological side, the limit on human intelligence up to now has been set by the size of the brain that will pass through the birth canal. Having watched my three children being born, I know how difficult it is for the head to get out. But within the next hundred years, I expect we will be able to grow babies outside the human body, so this limitation will be removed. Ultimately, however, increases in the size of the human brain through genetic engineering will come up against the problem that the body's chemical messengers responsible for our mental

Neural implants will offer enhanced memory and complete packages of information, such as an entire language or the contents of this book learned within minutes. Such enhanced humans will bear little resemblance to ourselves.

A BRIEF HISTORY OF THE UNIVERSE

EVENTS *(not to scale)*

0.00003 billion years.
The big bang and a fiery, optically dense, inflationary universe.

Matter/energy decouple.
The universe is transparent.

1 billion years.
Clusters of matter form protogalaxies synthesizing heavier nuclei.

3 billion years.
Galaxies recorded by Hubble Space Telescope in its Deep Field exploration.

TIME SEQUENCE *(to scale)*

0

1 billion

3 billion

5 billion

(FIG. 6.7)
The human race has been in existence for only a tiny fraction of the history of the universe. (If this chart was to scale and the length that human beings have been around was 7cm, then the whole history of the universe would be over a kilometer.) Any alien life we meet is likely to be much more primitive or much more advanced than we are.

activity are relatively slow-moving. This means that further increases in the complexity of the brain will be at the expense of speed. We can be quick-witted or very intelligent, but not both. Still, I think we can become a lot more intelligent than most of the people in *Star Trek*, not that that might be difficult.

Electronic circuits have the same complexity-versus-speed problem as the human brain. In this case, however, the signals are electrical, not chemical, and travel at the speed of light, which is much higher. Nevertheless, the speed of light is already a practical limit on the design of faster computers. One can improve the situation by making the circuits smaller, but ultimately there will be a limit set by the atomic nature of matter. Still, we have some way to go before we meet that barrier.

Another way in which electronic circuits can increase their

New galaxies, like our own, with heavier nuclei, are formed.

Formation of our solar system with orbiting planets.

3.5 billion years ago life-forms begin to appear.

0.0005 billion years ago early humans appear.

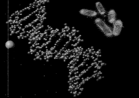

10.3 billion

11.5 billion

15 billion

complexity while maintaining speed is to copy the human brain. The brain does not have a single CPU—central processing unit—that processes each command in sequence. Rather, it has millions of processors working together at the same time. Such massively parallel processing will be the future for electronic intelligence as well.

Assuming we don't destroy ourselves in the next hundred years, it is likely that we will spread out first to the planets in the solar system and then to the nearby stars. But it won't be like *Star Trek* or *Babylon 5*, with a new race of nearly human beings in almost every stellar system. The human race has been in its present form for only two million years out of the fifteen billion years or so since the big bang (Fig. 6.7).

So even if life develops in other stellar systems, the chances of

THE BIOLOGICAL-ELECTRONIC INTERFACE

Within two decades a thousand-dollar computer may be as complex as the human brain. Parallel processors could mimic the way our brain works and make computers act in intelligent and conscious ways.

Neural implants may allow a much faster interface between the brain and computers, dissolving the distance between biological and electronic intelligence.

In the near future, most business transactions will probably be made between cyberpersonalities via the World Wide Web.

Within a decade, many of us may even choose to live a virtual existence on the Net, forming cyberfriendships and relationships.

Our understanding of the human genome will undoubtedly create great medical advances, but it will also enable us to increase the complexity of the human DNA structure significantly. In the next few hundred years, human genetic engineering may replace biological evolution, redesigning the human race and posing entirely new ethical questions.

Space travel beyond our solar system will probably require either genetically engineered humans or unmanned computer-controlled probes.

catching it at a recognizably human stage are very small. Any alien life we encounter will likely be either much more primitive or much more advanced. If it is more advanced, why hasn't it spread through the galaxy and visited Earth? If aliens had come here, it should have been obvious: more like the film *Independence Day* than *E.T.*

So how does one account for our lack of extraterrestrial visitors? It could be that there is an advanced race out there which is aware of our existence but is leaving us to stew in our own primitive juices. However, it is doubtful it would be so considerate to a lower life-form: do most of us worry how many insects and earthworms we squash underfoot? A more reasonable explanation is that there is a very low probability either of life developing on other planets or of that life developing intelligence. Because we claim to be intelligent, though perhaps without much ground, we tend to see intelligence as an inevitable consequence of evolution. However, one can question that. It is not clear that intelligence has much survival value. Bacteria do very well without intelligence and will survive us if our so-called intelligence causes us to wipe ourselves out in a nuclear war. So as we explore the galaxy we may find primitive life, but we are not likely to find beings like us.

The future of science won't be like the comforting picture painted in *Star Trek*: a universe populated by many humanoid races, with an advanced but essentially static science and technology. Instead, I think we will be on our own, but rapidly developing in biological and electronic complexity. Not much of this will happen in the next hundred years, which is all we can reliably predict. But by the end of the next millennium, if we get there, the difference from *Star Trek* will be fundamental.

Does intelligence have much long-term survival value?

CHAPTER 7

BRANE NEW WORLD

Do we live on a brane or are we just holograms?

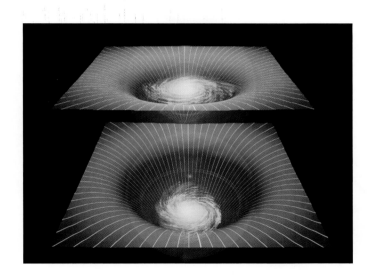

Here be dragons

(FIG. 7.1)

M-theory is like a jigsaw. It is easy to identify and fit together the pieces around the edges but we don't have much idea of what happens in the middle, where we can't make the approximation that some quantity or other will be small.

H OW WILL OUR JOURNEY OF DISCOVERY PROCEED IN THE future? Will we succeed in our quest for a complete unified theory that will govern the universe and everything that it contains? In fact, as described in Chapter 2, we may have already identified the Theory of Everything (ToE) as M-theory. This theory doesn't have a single formulation, at least as far as we know. Instead we have discovered a network of apparently different theories that all seem to be approximations to the same underlying fundamental theory in different limits, just as Newton's Theory of Gravity is an approximation to Einstein's General Theory of Relativity in the limit that the gravitational field is weak. M-theory is like a jigsaw: it is easiest to identify and fit together the pieces round the edges of the jigsaw, the limits of M-theory where some quantity or other is small. We now have a fairly good idea of these edges but there is still a gaping hole at the center of the M-theory jigsaw where we don't know what is going on (Fig. 7.1). We can't really claim to have found the Theory of Everything until we have filled that hole.

What is in the center of M-theory? Will we discover dragons (or something equally strange) like on old maps of unexplored lands? Our experience in the past suggests we are likely to find unexpected new phenomena whenever we extend the range of our observations to smaller scales. At the begining of the twentieth century, we understood the workings of nature on the scales of classical physics, which is good from interstellar distances down to about a hundredth of a millimeter. Classical physics assumes that matter is a

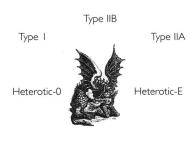

Type IIB

Type I   Type IIA

Heterotic-0   Heterotic-E

11-dimensional supergravity

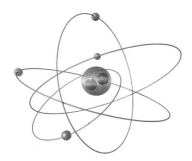

(FIG. 7.2)
Right: The classical indivisible atom.
Far right: An atom showing electrons orbiting a nucleus of protons and neutrons.

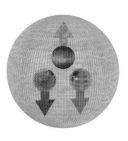

(FIG. 7.3)
Top: A proton consists of two up quarks, each with a positive two-thirds electrical charge, and one down quark, having a negative one-third charge. *Bottom:* A neutron consists of two down quarks, each with a negative one-third electrical charge, and one up quark, having a positive two-thirds charge.

continuous medium with properties like elasticity and viscosity, but evidence began to emerge that matter is not smooth but grainy: it is made of tiny building blocks called atoms. The word atom comes from Greek and means indivisible, but it was soon found that atoms consisted of electrons orbiting a nucleus made up of protons and neutrons (Fig. 7.2).

The work on atomic physics in the first thirty years of the century took our understanding down to lengths of a millionth of a millimeter. Then we discovered that protons and neutrons are made of even smaller particles called quarks (Fig. 7.3).

Our recent research on nuclear and high-energy physics has taken us to length scales that are smaller by a further factor of a billion. It might seem that we could go on forever, discovering structures on smaller and smaller length scales. However, there is a limit to this series, as there is to the series of Russian dolls within Russian dolls (Fig. 7.4).

Eventually, one gets down to a smallest doll, which can't be taken apart any more. In physics, the smallest doll is called the Planck length. To probe to shorter distances would require particles of such high energy that they would be inside black holes. We don't know exactly what the fundamental Planck length is in M-theory, but it might be as small as a millimeter divided by a hundred thousand billion billion billion. We are not about to build particle accelerators that can probe to distances that small. They would have to

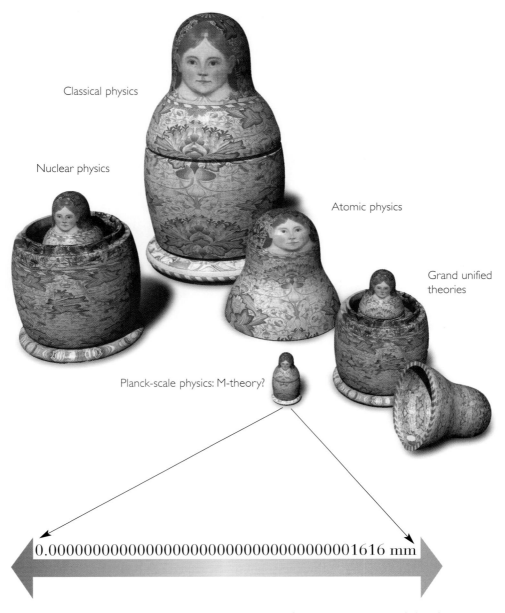

Classical physics

Nuclear physics

Atomic physics

Grand unified theories

Planck-scale physics: M-theory?

0.0000000000000000000000000000000001616 mm

(FIG. 7.4) Each doll represents a theoretical understanding of nature down to a certain length scale. Each contains a smaller doll that corresponds to a theory that describes nature on shorter scales. But there exists a smallest fundamental length in physics, the Planck length, a scale at which nature may be described by M-theory.

(FIG. 7.5)

The size of an accelerator needed to probe distances as small as a Planck length would be greater than the diameter of the solar system.

be larger than the solar system, and they are not likely to be approved in the present financial climate (Fig. 7.5).

However, there has been an exciting new development that means we might discover at least some of the dragons of M-theory more easily (and cheaply). As explained in Chapters 2 and 3, in the M-theory network of mathematical models, spacetime has ten or eleven dimensions. Up to recently it was thought that the six or seven extra dimensions would all be curled up very small. It would be like a human hair (Fig.7.6).

If you look at a hair under a magnifying glass, you can see it has thickness, but to the naked eye it just appears like a line with length but no other dimension. Spacetime may be similar: on human, atomic, or even nuclear physics length scales, it may appear four-dimensional and nearly flat. On the other hand if we probe to very short distances using extremely high energy particles, we should see that spacetime was ten- or eleven-dimensional.

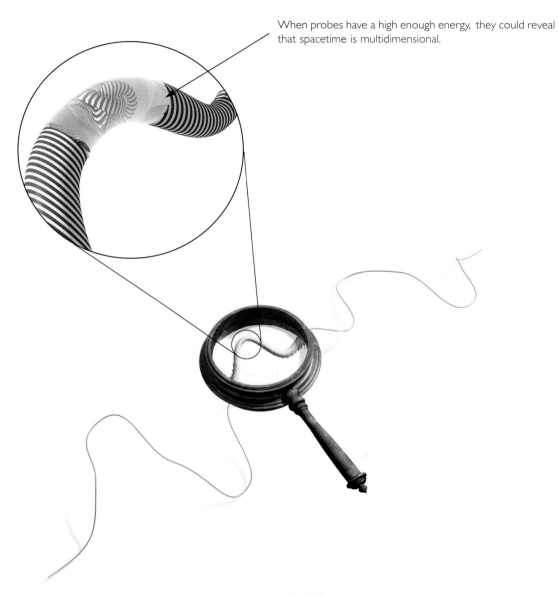

When probes have a high enough energy, they could reveal that spacetime is multidimensional.

(FIG. 7.6)
To the naked eye a hair looks like a line; its only dimension appears to be length. Similarly, spacetime may look four-dimensional to us, but appear ten- or eleven-dimensional when probed with very high-energy particles.

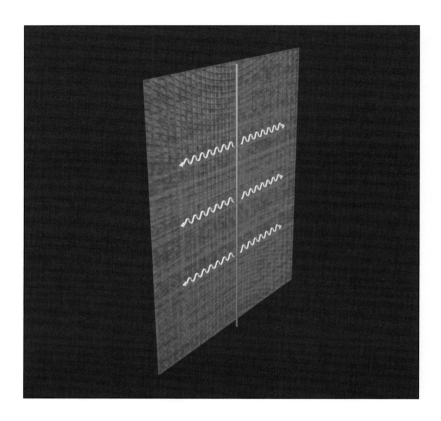

(FIG. 7.7) BRANE WORLDS

The electric force would be confined to the brane and would fall off at the right rate for electrons to have stable orbits about the nuclei of atoms.

If all the additional dimensions were very small, it would be very difficult to observe them. However, there has recently been the suggestion that one or more of the extra dimensions might be comparitively large or even infinite. This idea has the great advantage (at least to a positivist like me) that it may be testable by the next generation of particle accelerators or by sensitive short-range measurements of the gravitational force. Such observations could either falsify the theory or experimentally confirm the presence of other dimensions.

Large extra dimensions are an exciting new development in our search for the ultimate model or theory. They would imply that we lived in a brane world, a four-dimensional surface or brane in a higher-dimensional spacetime.

Matter and nongravitational forces like the electric force would be confined to the brane. Thus everything not involving gravity would

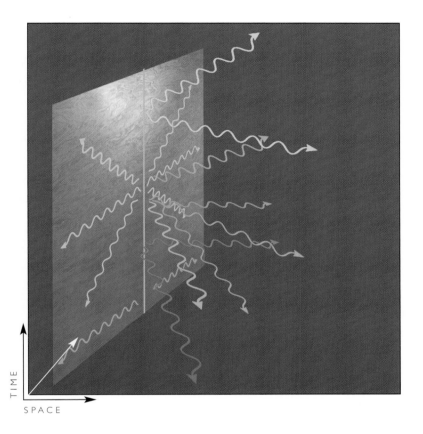

TIME

SPACE

behave as it would in four dimensions. In particular, the electric force between the nucleus of an atom and the electrons orbiting around it would fall off with distance at the right rate for atoms to be stable against the electrons falling into the nucleus (Fig. 7.7).

This would be in accordance with the anthropic principle that the universe must be suitable for intelligent life: if atoms weren't stable, we wouldn't be here to observe the universe and ask why it appears four-dimensional.

On the other hand, gravity in the form of curved space would permeate the whole bulk of the higher-dimensional spacetime. This would mean that gravity would behave differently from other forces we experience: because gravity would spread out in the extra dimensions, it would fall off more rapidly with distance than one would expect (Fig. 7.8).

(FIG. 7.8)
Gravity would spread into the extra dimensions as well as acting along the brane, and would fall off faster with distance than it would in four dimensions.

181

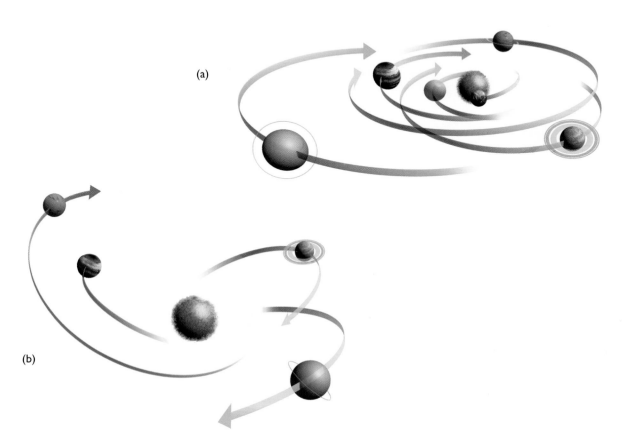

(a)

(b)

(FIG. 7.9)
A faster falloff of the gravitational force at large distances would mean that planetary orbits would be unstable. Planets would either fall into the Sun (a) or escape its attraction altogether (b).

If this more rapid falloff of the gravitational force extended to astronomical distances, we would have noticed its effect on the orbits of the planets. In fact they would be unstable, as was remarked in Chapter 3: the planets would either fall into the Sun or escape to the dark and cold of interstellar space (Fig. 7.9).

However, this would not happen if the extra dimensions ended on another brane not that far away from the brane on which we live. Then for distances greater than the separation of the branes, gravity would not be able to spread out freely but would effectively be confined to the brane, like the electric forces, and fall off at the right rate for planetary orbits (Fig. 7.10).

On the other hand, for distances less than the separation of the branes, gravity would vary more rapidly. The very small gravitational force between heavy objects has been measured accurately in

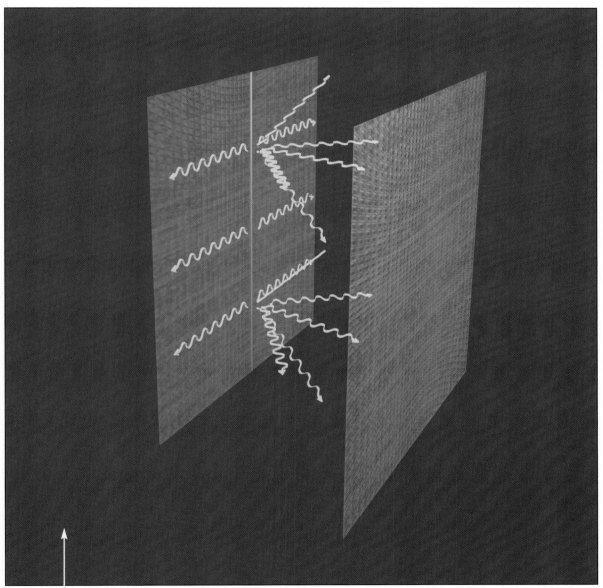

Extra dimensions

(FIG. 7.10) A second brane near our brane world would prevent gravity from spreading far into the extra dimensions and would mean that at distances greater than the brane separation, gravity would fall off at the rate one would expect for four dimensions.

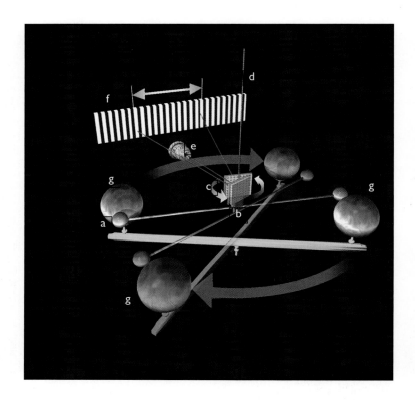

(FIG. 7.11)

THE CAVENDISH EXPERIMENT

A laser beam (e) determines any twist of the dumbbell as it is projected on a calibrated screen (f). Two small lead spheres (a) attached to the dumbbell (b) with a small mirror (c) are freely suspended by a torsion fiber.

Two large lead spheres (g) are placed near the small ones on a rotating bar. As the larger lead spheres rotate to the opposite position, the dumbbell oscillates and then settles to a new position.

the lab but the experiments so far would not have detected the effects of branes separated by less than a few millimeters. New measurements are now being made at shorter distances (Fig. 7.11).

In this brane world, we would live on one brane but there would be another "shadow" brane nearby. Because light would be confined to the branes and would not propagate through the space between, we could not see the shadow world. But we would feel the gravitational influence of matter on the shadow brane. In our brane such gravitational forces would appear to be produced by sources that were truly "dark" in that the only way we could detect them is through their gravity (Fig. 7.12). In fact in order to explain the rate at which stars orbit the center of our galaxy, it seems there must be more mass than is accounted for by the matter we observe.